Arduino™ + Android™ Projects for the Evil Genius™

Evil Genius™ Series

Arduino + Android Projects for the Evil Genius
Bike, Scooter, and Chopper Projects for the Evil Genius
Bionics for the Evil Genius: 25 Build-it-Yourself Projects
Electronic Circuits for the Evil Genius, Second Edition: 64 Lessons with Projects
Electronic Gadgets for the Evil Genius: 28 Build-it-Yourself Projects
Electronic Sensors for the Evil Genius: 54 Electrifying Projects
15 Dangerously Mad Projects for the Evil Genius
50 Awesome Auto Projects for the Evil Genius
50 Green Projects for the Evil Genius
50 Model Rocket Projects for the Evil Genius
51 High-Tech Practical Jokes for the Evil Genius
46 Science Fair Projects for the Evil Genius
Fuel Cell Projects for the Evil Genius
Holography Projects for the Evil Genius
Mechatronics for the Evil Genius: 25 Build-it-Yourself Projects
Mind Performance Projects for the Evil Genius: 19 Brain-Bending Bio Hacks
MORE Electronic Gadgets for the Evil Genius: 40 NEW Build-it-Yourself Projects
101 Outer Space Projects for the Evil Genius
101 Spy Gadgets for the Evil Genius, Second Edition
123 PIC® Microcontroller Experiments for the Evil Genius
123 Robotics Experiments for the Evil Genius
125 Physics Projects for the Evil Genius
PC Mods for the Evil Genius: 25 Custom Builds to Turbocharge Your Computer
PICAXE Microcontroller Projects for the Evil Genius
Programming Video Games for the Evil Genius
Recycling Projects for the Evil Genius
Solar Energy Projects for the Evil Genius
Telephone Projects for the Evil Genius
30 Arduino Projects for the Evil Genius
TinyAVR Microcontroller Projects for the Evil Genius
22 Radio and Receiver Projects for the Evil Genius
25 Home Automation Projects for the Evil Genius

Arduino™ + Android™ Projects for the Evil Genius™

Control Arduino with Your Smartphone or Tablet

Simon Monk

New York Chicago San Francisco Lisbon London Madrid
Mexico City Milan New Delhi San Juan Seoul
Singapore Sydney Toronto

The McGraw-Hill Companies

Cataloging-in-Publication Data is on file with the Library of Congress

McGraw-Hill books are available at special quantity discounts to use as premiums and sales promotions, or for use in corporate training programs. To contact a representative, please e-mail us at bulksales@mcgraw-hill.com.

Arduino™ + Android™ for the Evil Genius™: Control Arduino with Your Smartphone or Tablet

1 2 3 4 5 6 7 8 9 QDB QDB 1 0 9 8 7 6 5 4 3 2 1

ISBN 978-0-07-177596-0
MHID 0-07-177596-X

Sponsoring Editor
Roger Stewart

Editorial Supervisor
Jody McKenzie

Acquisitions Coordinator
Joya Anthony

Project Manager
Patricia Wallenburg

Copy Editor
Mike McGee

Proofreader
Claire Splan

Indexer
Claire Splan

Production Supervisor
George Anderson

Composition
TypeWriting

Art Director, Cover
Jeff Weeks

Cover Designer
Todd Radom

To Linda. The love of my life.

About the Author

Simon Monk has a bachelor's degree in Cybernetics and Computer Science and a doctorate in Software Engineering. He has been an active electronics hobbyist since his school days and is an occasional author in hobby electronics magazines. He is also author of *30 Arduino Projects for the Evil Genius* and *15 Dangerously Mad Projects for the Evil Genius*.

Contents at a Glance

PART ONE Android Peripherals

1 Bluetooth Robot . . . 3

2 Android Geiger Counter . . . 17

3 Android Light Show . . . 37

4 TV Remote . . . 55

5 Temperature Logger . . . 63

6 Ultrasonic Range Finder . . . 73

PART TWO Home Automation

7 Home Automation Controller . . . 85

8 Power Control . . . 111

9 Smart Thermostat . . . 129

10 RFID Door Lock . . . 145

11 Signaling Flags . . . 163

12 Delay Timer . . . 171

Appendix: Open Accessory Primer . . . 183

Index . . . 193

Contents

Acknowledgments . . . xiii

Introduction . . . xv

PART ONE Android Peripherals

1 Bluetooth Robot . . . 3
Construction . . . 3
Theory . . . 14
Summary . . . 15

2 Android Geiger Counter . . . 17
Google Open Accessory . . . 18
Construction . . . 18
Theory . . . 30
Summary . . . 35

3 Android Light Show . . . 37
Construction: The Droid Accessory Base . . . 38
Construction: The Light Show Project . . . 43
Using the Project . . . 50
Theory . . . 50
Summary . . . 53

4 TV Remote . . . 55
Construction . . . 56
Using the Project . . . 60
Theory . . . 60
Summary . . . 61

5 Temperature Logger . . . 63
Construction . . . 64
Using the Project . . . 68
Theory . . . 70
Summary . . . 71

6 Ultrasonic Range Finder . . . 73
Construction . . . 74
Using the Project . . . 79
Theory . . . 79
Summary . . . 81

PART TWO Home Automation

7 **Home Automation Controller** **85**
The Sound Link Module 87
Android Software 99
Internet Access 102
Theory 103
Summary 110

8 **Power Control** **111**
Power Control Electronics 111
Constructing the Power Control Module 112
Adding It to the Home Automation Controller 120
Setting Up Your Home 124
Theory 124
Summary 128

9 **Smart Thermostat** **129**
Construction 130
Using the System 140
Theory 141
Summary 144

10 **RFID Door Lock** **145**
Construction 146
Using the System 156
Theory 157
Summary 161

11 **Signaling Flags** **163**
Construction 164
Theory 168
Summary 169

12 **Delay Timer** **171**
Construction 171
Theory 178
Summary 181

Appendix: Open Accessory Primer **183**
Learning Android Programming 183
Arduino Programming 183
The Example 183
On the Arduino 184
Android 186
Conclusion 192

Index 193

Acknowledgments

I THANK LINDA for giving me the time, space, and support to write this book and for putting up with the various messes my projects create around the house.

I also thank my boys, Stephen and Matthew Monk, for taking an interest in what their Dad is up to and their general assistance with project work.

Finally, I would like to thank Roger Stewart, Patricia Wallenburg, Mike McGee, and everyone at McGraw-Hill, who did a great job once again. It's a pleasure to work with such a great team.

Introduction

This is a project book that marries together the simple-to-use microcontroller boards (Arduino) and the world of Android mobile phones and tablet computers.

The book contains detailed instructions for constructing various projects that use Arduino and Android devices. Some of the projects, such as the Geiger Counter and the Ultrasonic Distance Meter, are essentially electronic accessories for your Android phone.

Other projects in the book work toward a home automation system, complete with electric door lock and a remote control for power and heating, making even home automation accessible from the Internet and your Android device.

Arduino

Arduino (Figure 1) is a small microcontroller board with a USB plug to connect to your computer and a number of connection sockets that can be wired up to external electronics such as motors, relays, light sensors, laser diodes, loudspeakers, microphones, and other items. They can either be powered through the USB connection from the computer, or from a battery or other

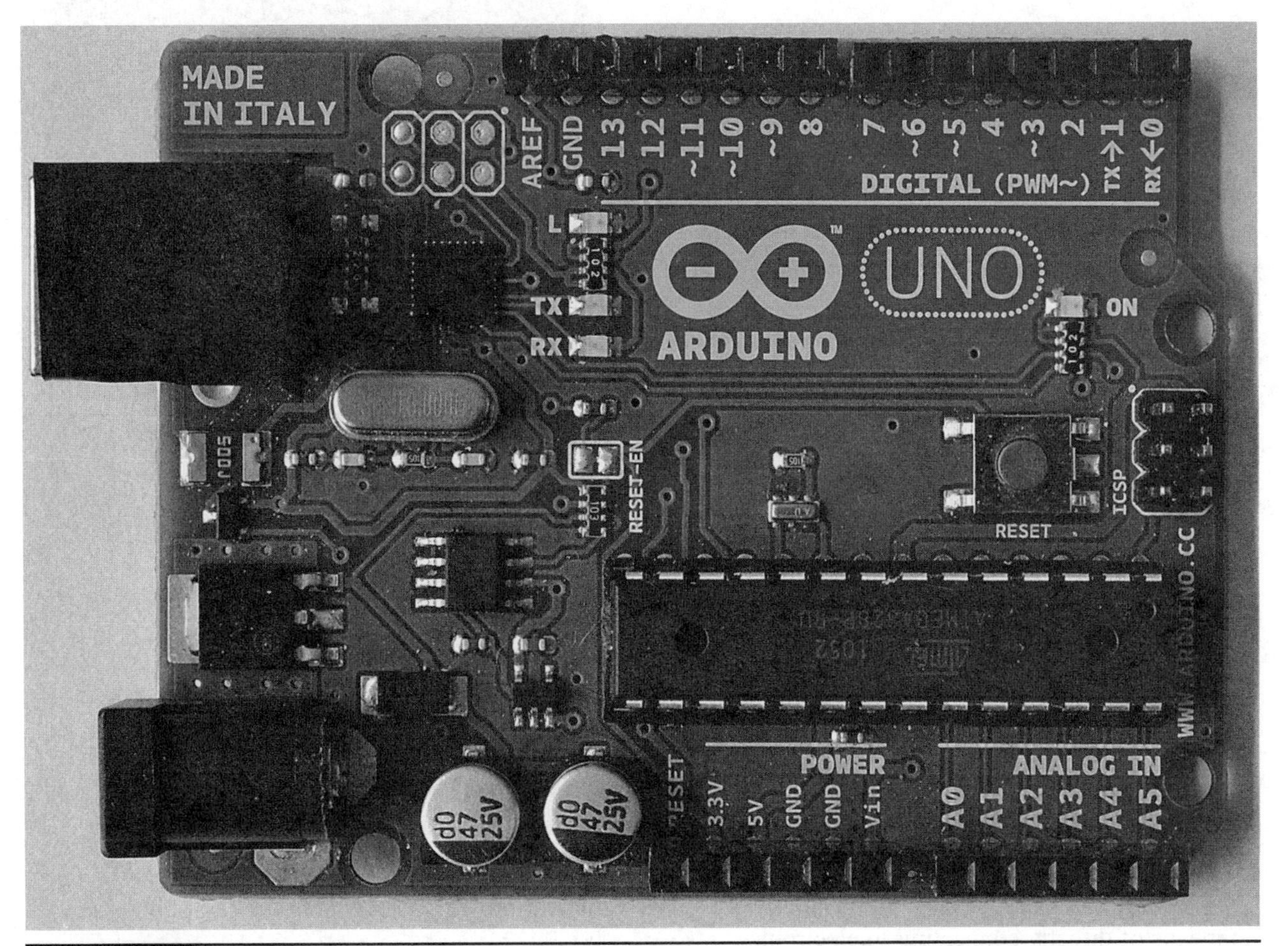

Figure 1 An Arduino Uno board

power supply. They can be controlled from the computer or programmed by the computer and then disconnected and allowed to work independently.

The board design is open source. This means that anyone is allowed to make Arduino-compatible boards. Such competition has led to low costs.

The basic boards are supplemented by accessory shield boards that can be plugged on top of the Arduino board. In this book, we will use three shields: the USB master shield that allows us to connect to Android devices over USB; a motor shield for driving the wheels of a little robot; and an Ethernet shield that will allow us to turn our Arduino into a tiny web server.

The software for programming your Arduino is easy to use and also freely available for Windows, Mac, and Linux computers, at no cost.

Android

Android is Google's phone and tablet operating system. Developing for Android is free. The software development tools are free and there are no fees associated with distributing your app. You can also deploy directly without having to use Google's Market.

Apps for all the Android projects in the book, such as the one shown in Figure 2, are available for download from the book's web site at www.duinodroid.com. However, should you wish to modify the apps, the source code is also freely available from the web site.

Arduino and Android

Arduino is all about electronics connecting things together, but it lacks much in the way of a user interface and wireless connectivity. Android, on the other hand, has great user interface capabilities but no means of directly connecting to electronics.

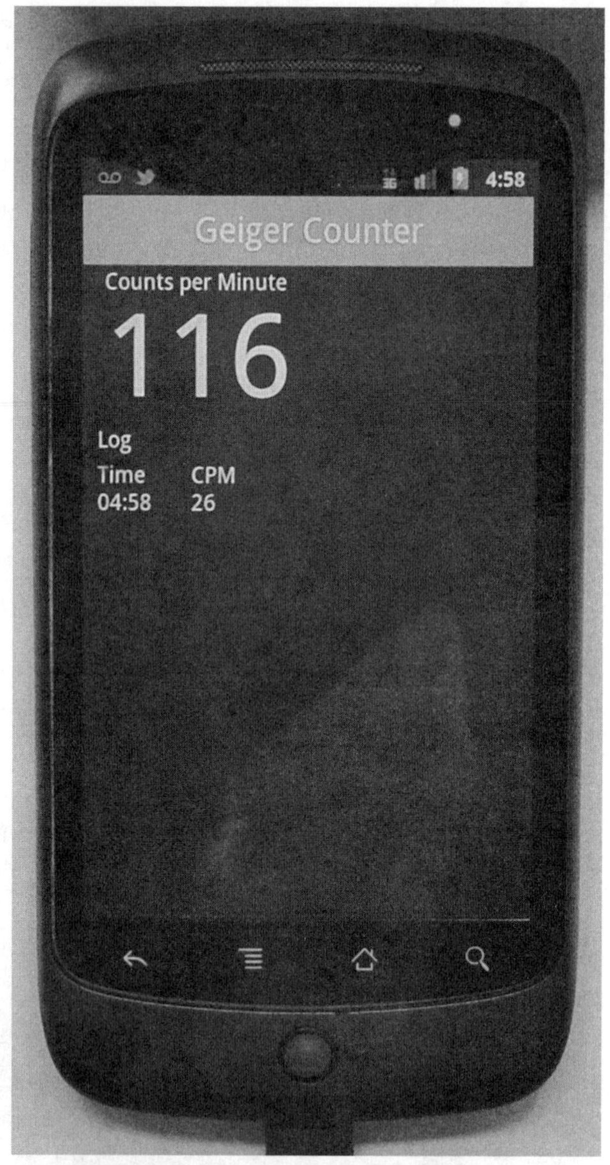

Figure 2 The Geiger Counter app

Putting these two together, the Evil Genius can do great things!

Android Open Accessory

At Google's developer conference (Google IO 2011), the Open Accessory standard was announced. This is the standard for creating hardware accessories for Android devices that plug into the Android device through its USB connection. This is available for cell phones and tablets with Android 2.3.4 or later.

The really cool thing about this standard is that it is based around Arduino technology. Great news for Arduino enthusiasts and five of the projects in this book (Geiger Counter, Light Show, TV Remote, Temperature Logger, and Ultrasonic Range Finder) are Open Accessory projects.

The book introduces the concept of a Droid Duino Base, which takes an Arduino's microcontroller off the Arduino board and fits it into a socket on the prototyping area of a USB host shield. This removes the need for an Arduino for each project, reducing the size and keeping the cost down to a few tens of dollars per project. This forms the basis of four of the projects in the book and makes a handy module to reuse in your own Open Accessory projects.

Amarino

In some situations, a wired connection is not what is needed. For example, in the first project in this book (Bluetooth Robot), wires would really cramp the robot's style. We can, however, use a clever bit of technology called Amarino that will allow us to remote control a small robot from an app on an Android phone.

Sound Interface

The second section of the book is concerned with building a home automation system using a low-cost Android tablet as the controller, which communicates with an Arduino device to provide the electronic interface. These tablets do not often have Bluetooth or Open Accessory capability, so a wired interface between the tablet and an Arduino is made using the audio jack.

This uses the same approach as the cassette tape interfaces employed by home computers in the 1980s.

The Book

All the projects in this book contain step-by-step construction details. All require some soldering, so a basic familiarity with soldering is required.

Schematic diagrams and layouts for stripboard or perfboard are provided.

All the Arduino sketches and Android apps are made freely available, so you do not need to know how to program. However, the software is explained for those who want to modify the designs or understand the basic principals before designing their own projects.

The book also includes an Android Open Accessory Primer appendix for those wishing to understand more about this framework and how to program it on both the Arduino and Android sides.

Projects

Each of the projects in this book is contained in its own chapter. Most of the projects can be built in isolation; however, the home automation projects of Chapters 8, 9, 10, and 11 all require the home automation controller of Chapter 7 to be built.

The projects in this book are summarized in the following table on the next page.

The number of stars under the Difficulty column for each project will give you an idea of the ease of construction. The more stars, the more difficult the project. None of the projects require any surface-mount soldering or indeed anything finer than a 1/10-inch-pitch stripboard.

Chapter	Project	Notes	Difficulty
1	Bluetooth Robot	Control a small rover from your Android phone using Bluetooth and an Arduino and motor shield.	★★
2	Android Geiger Counter	An Android Open Accessory project using a USB host shield and Arduino Uno.	★★★★
3	Android Light Show	A powered Open Accessory dock for your Android phone that drives three LED panels to make a sound-sensitive light show.	★★★
4	Android TV Remote	Programmable infrared remote control accessory for your Android phone.	★★★
5	Temperature Logger	A temperature logger that uses the Android phone to wirelessly send readings to Pachube.	★★★
6	Ultrasonic Range Finder	An Android Open Accessory project for measuring distance.	★★★
7	Home Automation Controller	The Android tablet base unit with an audio jack interface to Arduino.	★★★
8	Home Automation Power Control	Add control of AC outlets and lights to the Home Automation Controller project from Chapter 7.	★★
9	Home Automation Thermostat	Add in remote control of domestic heating to the home automation controller using low-cost RF data modules.	★★★
10	Remote Door Lock	Control access to your home using RFID tags; also includes a RF link to the home automation controller.	★★★
11	Ethernet Flag	Network control of two flags that can be activated from any Internet device. Useful for the Evil Genius to summon minions.	★★
12	Delay Timer	A simple-to-make Arduino-based delay timer.	★★★

Components

All the components used are readily available and suppliers as well as part numbers are given wherever appropriate. Farnell part numbers are provided for the standard components. Even if you do not order them from Farnell, this can be helpful to identify exactly what the component is before ordering elsewhere.

SparkFun is a quick and reliable supplier of Arduino-related hardware, and in the UK, Proto-PIC has a good range of Arduino-related boards and shields at competitive prices.

For other items, eBay is always a good low-cost source of components, but quality can vary.

Getting Started

If you are interested in the Android Open Accessory projects then the Android Light Show project is not a bad project to begin with. It includes the instructions for building the Droid Duino Base that is the main component of the other Open Accessory projects (with the exception of the Geiger Counter).

For those budding Evil Geniuses more interested in automating the Evil Genius Lair, start with Chapter 7, the Home Automation Controller, as this is the basis for the subsequent home automation projects.

If you find yourself wanting to know more about using the Arduino, you may wish to look at the other books by this author: *Programming Arduino* and *30 Arduino Projects for the Evil Genius*.

For source code, build apps, and much more, please visit the book's web site at www.duinodroid.com.

Arduino™ + Android™ Projects for the Evil Genius™

PART ONE

Android Peripherals

CHAPTER 1

Bluetooth Robot

The Evil Genius and his Android phone are inseparable. You will find him using it for grocery shopping, redirecting military satellites, and messaging his minions. The Evil Genius also loves to direct small robots by remote control using Bluetooth.

This project employs a simple Android app (Figure 1-1) and an Arduino-controlled robot using a low-cost Bluetooth module (Figure 1-2).

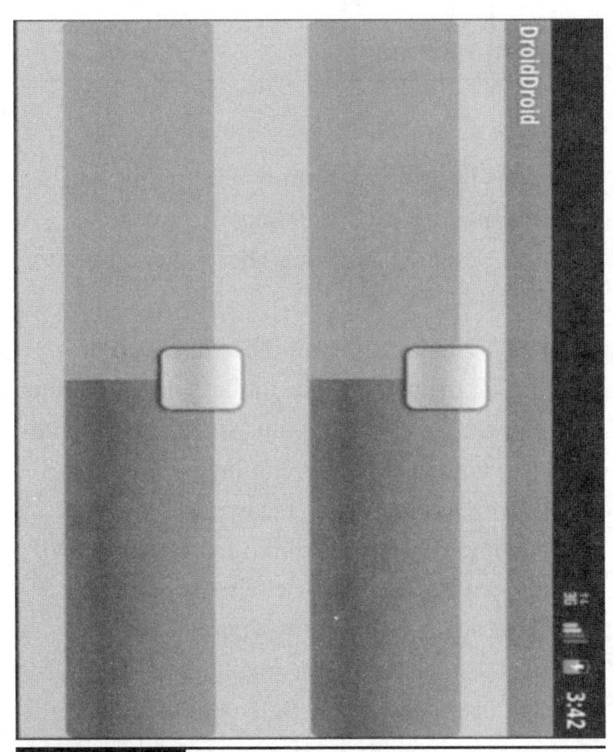

Figure 1-1 A remote-control app

Figure 1-2 A Bluetooth robot

Arduino boards are very popular microcontroller boards that have a number of advantages, not the least of which are:

- They are easy to program, and can be programmed from a Mac, Windows, or Linux computer.
- Many "shields" just plug into the top of the Arduino board.
- They are not expensive.

Construction

Figure 1-3 shows the schematic diagram for the project.

The robot's drive motors are controlled by a motor shield kit, and the Bluetooth module is attached to the prototyping area of the shield, making this a simple project to construct, with a minimal amount of soldering required.

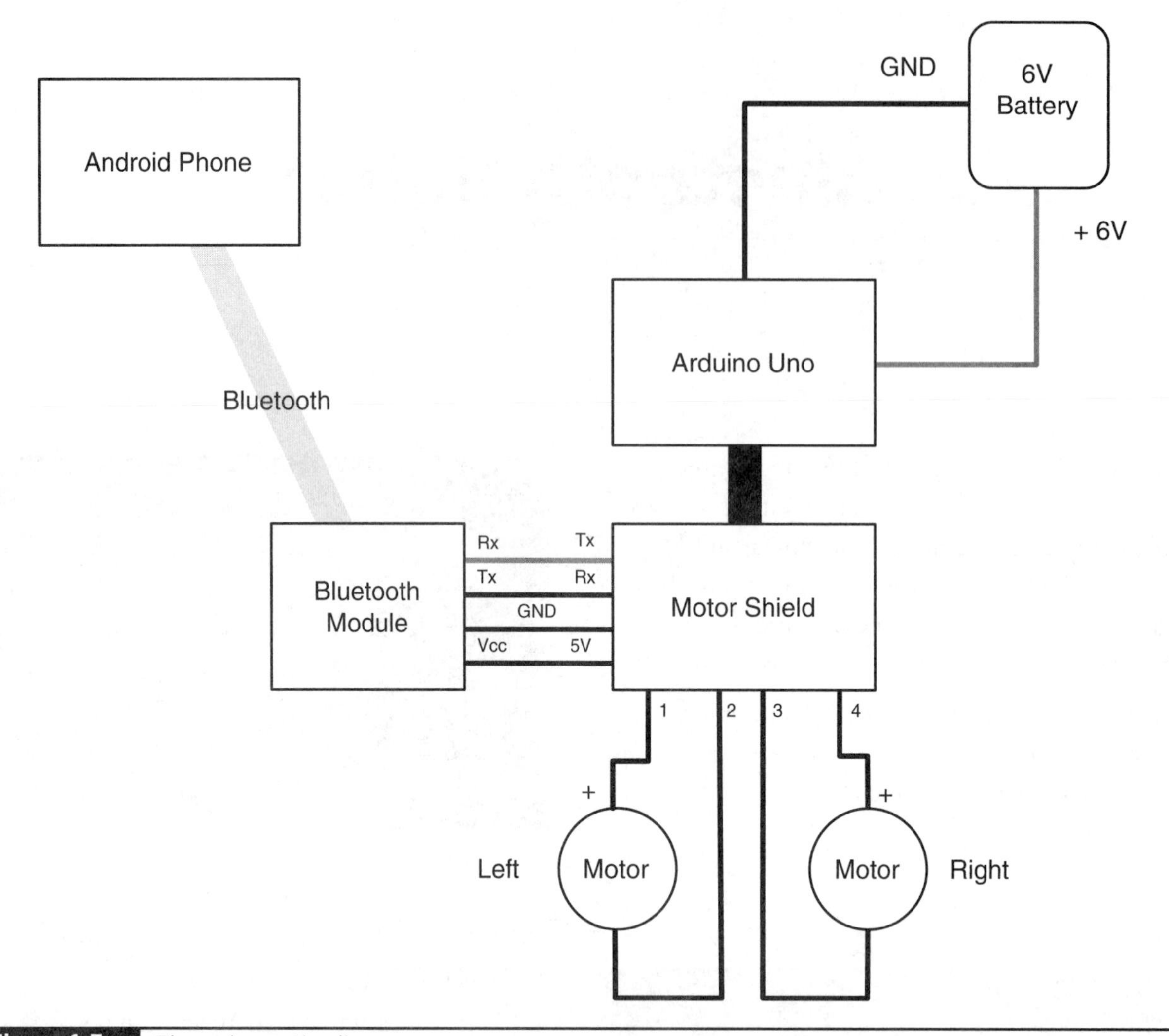

Figure 1-3 The schematic diagram

All the software for the project can be downloaded from www.duinodroid.com.

What You Will Need

In addition to a Bluetooth-equipped Android phone (Android 2.1 or later), you will need the components found in the following Parts Bin to make the project.

This design uses the Arduino Uno. The official Arduino web site (www.arduino.cc) lists suppliers of the Uno. However, if you are on a budget, you can use a clone of the Arduino Uno. The Arduino is "open-source hardware," which means all the design files are available under a Creative Commons license that permits third parties to make their own Arduinos. Many do, and an Internet search will find you cheap alternatives to the official "Uno."

There are many different types of Bluetooth modules on the market. The module that the author used is a simple "stick of gum"–shaped Bluetooth module with just four pins coming out of it that supply power and provide, receive, and transmit connections. These types of modules work at 5V and are ideally suited for use with an Arduino. They are usually made up of a base board with four pins on which an even smaller board is mounted that is the actual Bluetooth module. These can be bought on eBay for around USD 15.

PARTS BIN

Part	Quantity	Description	Source
Arduino Uno	1	Arduino Uno board	www.arduino.cc
Ardumoto	1	Ardumoto Motor Shield	Sparkfun: DEV-09815
BT Module	1	TTL Bluetooth module: Bluesmirf or the equivalent	eBay, Sparkfun
Pin headers	1	Pin header strip, broken into two sections of six pins, and two sections of eight pins	Farnell: 1097954
Screw terminals	3	2-way screw terminal, 3.5mm pitch	Farnell: 1217302
Gear motors	2	120:1 mini plastic gear motor	Pololu: 1125
Switch	1	SPST miniature toggle switch	Farnell: 1661841
Battery holder	1	AAA battery holder, solder tags (4x)	Farnell: 1650687
Case	1	Plastic case, 145 × 80 × 30mm	
Wheels	2	Approximately 2-inch- (50mm-) diameter model car wheels	Model/toy shop
Caster	1	Small caster	Hardware store

It is best to buy one with the miniboard already soldered to the main board, because the connections are very tiny and quite hard to solder. Higher-quality and more expensive versions are available in the Bluesmirf range from suppliers like Sparkfun. The main difference between the low-cost and more expensive modules is range.

The gear motors from Pololu are ideal for this kind of application. They have about the right gear ratio and are not expensive. Alternatives are available, but try not to buy gear motors that draw more than about 1 amp, otherwise the motor shield will struggle to provide the necessary current.

The motor shield greatly simplifies the whole process of driving motors, and what's more, it has a handy little prototyping area at one end where you can attach your own extra components. In this case, that is where our Bluetooth module will sit. In the parts list, I specified the basic shield kit, which comes without header strips and screw terminals. This shield is also available as a "retail" kit that includes the pin header and screw terminals. See Sparkfun's web site for details.

In addition to these components, you will also need the following tools.

TOOLBOX

- An electric drill and assorted drill bits
- A hacksaw or Dremel rotary tool
- A hot glue gun or epoxy glue
- A computer to program the Arduino
- A USB-type A-to-B lead

Step 1. Attach Pin Headers to the Shield

The first step is to attach the pin headers to the motor shield. Figure 1-4 shows the bottom of the shield with the pin headers attached. Your header strips will probably come in a single long length that is designed to be snapped into sections of the

Figure 1-4 The motor shield with pin headers attached

Figure 1-5 The top side of the motor shield

correct length. You will need to break off two lengths of six pins and two lengths of eight pins.

The easiest way to keep the pins straight is to plug the long ends of the headers into your Arduino board while you solder them to the shield. However, this will heat up the plastic of the socket underneath and may distort it. So either solder quickly, or just solder the pins at each end of a section so the header is held in the right place, and then remove the shield and solder the rest of the pins.

When all the pins are in place, the top of the shield should look like Figure 1-5.

Figure 1-6 The shield with screw terminals attached

Step 2. Attach Screw Terminals to the Shield

The screw terminals fit next to the A and B channels for the motors. We will also attach a screw terminal to the power socket, as it is easier to connect to than the main 2.1mm power socket on the Arduino.

Solder the four screw terminals into place, with the opening facing toward the outside of the shield. Figure 1-6 shows the shield with the screw terminals attached and the shield plugged into an Arduino.

Step 3. Install the Bluetooth Module

Figure 1-7 shows the Bluetooth module soldered into place and the wiring attached.

Before attaching the Bluetooth module, carefully bend the pins with pliers so the board lies flat against the shield. Solder the module itself into place first, and then attach the four wires as listed next:

- +5V on the Bluetooth module to +5V on the shield
- GND on the Bluetooth module to GND on the shield

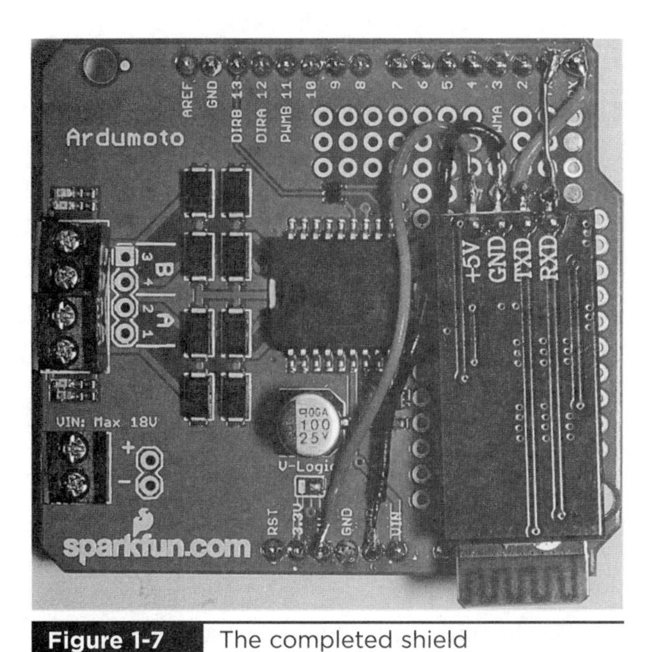

Figure 1-7 The completed shield

- TXD on the Bluetooth module to RX on the shield
- RXD on the Bluetooth module to TX on the shield

Note the cross-over between transmit and receive between the Arduino and the Bluetooth module.

That's it for the electronics. We now turn to constructing the hardware for the robot.

Step 4. Fix the Motors and Battery Box to the Case

Figure 1-8 shows the position of the motors. The plastic housing of the gear motors is glued to the inside of the box.

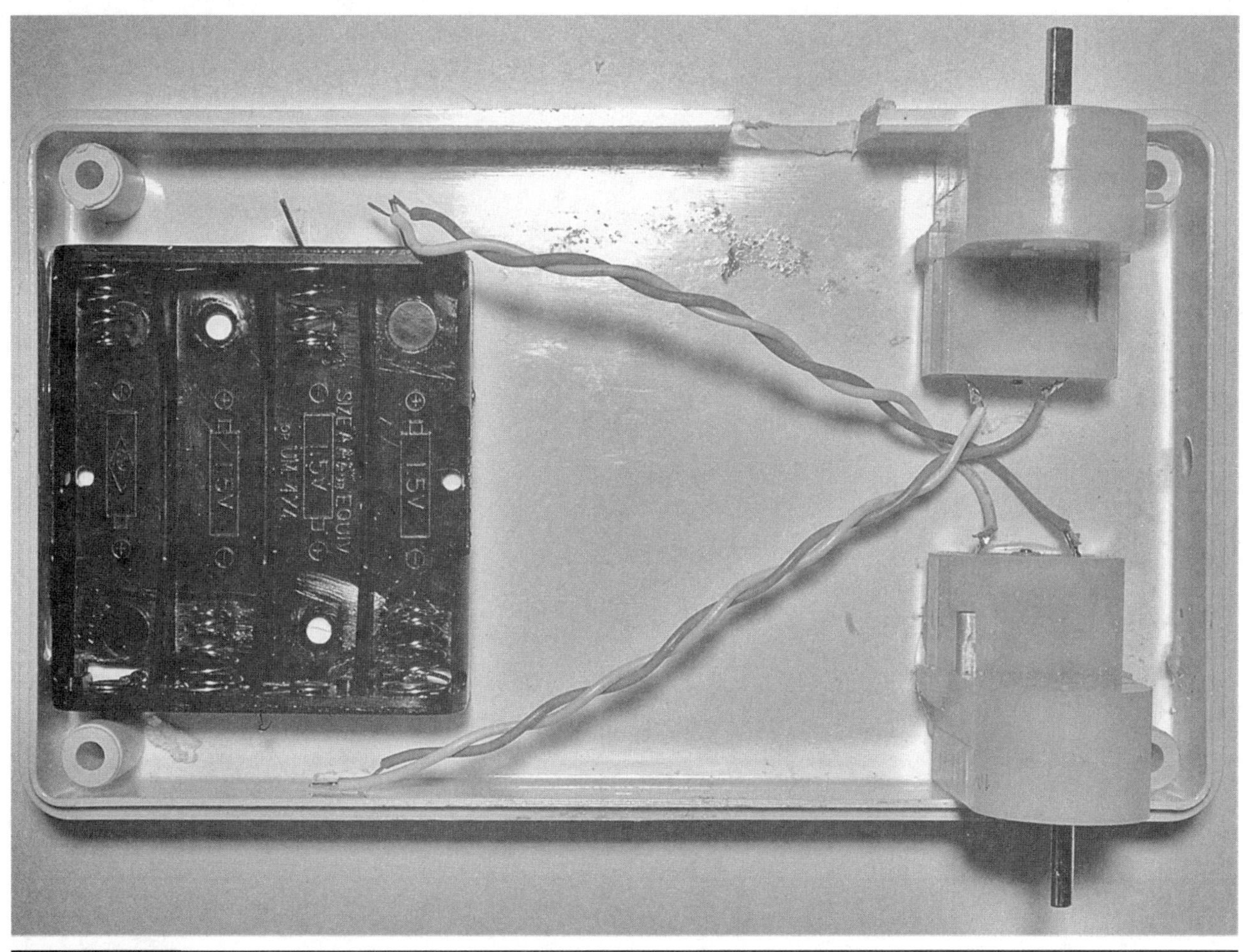

Figure 1-8 The motors and battery box glued to the inside of the box

The motors and battery box are fixed to the bottom of the case, at either end, leaving room for the Arduino and shield in the middle.

Step 5. Cut the Case Bottom and Fix the Castor

Figure 1-9 shows how the case bottom is cut to allow the top of the gear motors and their drive shafts to protrude from the bottom of the case.

It is also a good idea to make a hole close to the Bluetooth module so you can see if the LED on the module is flashing or solid. The box used by the author was reused from a previous project and had various holes in it. This is not a bad thing as it allows ventilation.

The castor is merely the smallest castor the author could find at the local hardware store. This is just glued to the bottom part of the case.

Step 6. Final Wiring

The wiring is made easy by the screw terminals. Figure 1-10 shows the wiring diagram, while Figure 1-11 displays a photo of the inside of the robot.

The wiring steps are:

1. Solder leads from each of the tabs on the gear motors. The leads should be long enough to comfortably reach the screw terminals on the motor shield.
2. Solder a lead directly from the negative power terminal on the shield to the negative of the battery holder.
3. Solder a longer lead to the center connection on the switch to the positive battery terminal.
4. Solder a shorter lead to one side of the switch (it does not matter which) and fit the other end into the positive screw terminal.

Figure 1-9 The case bottom

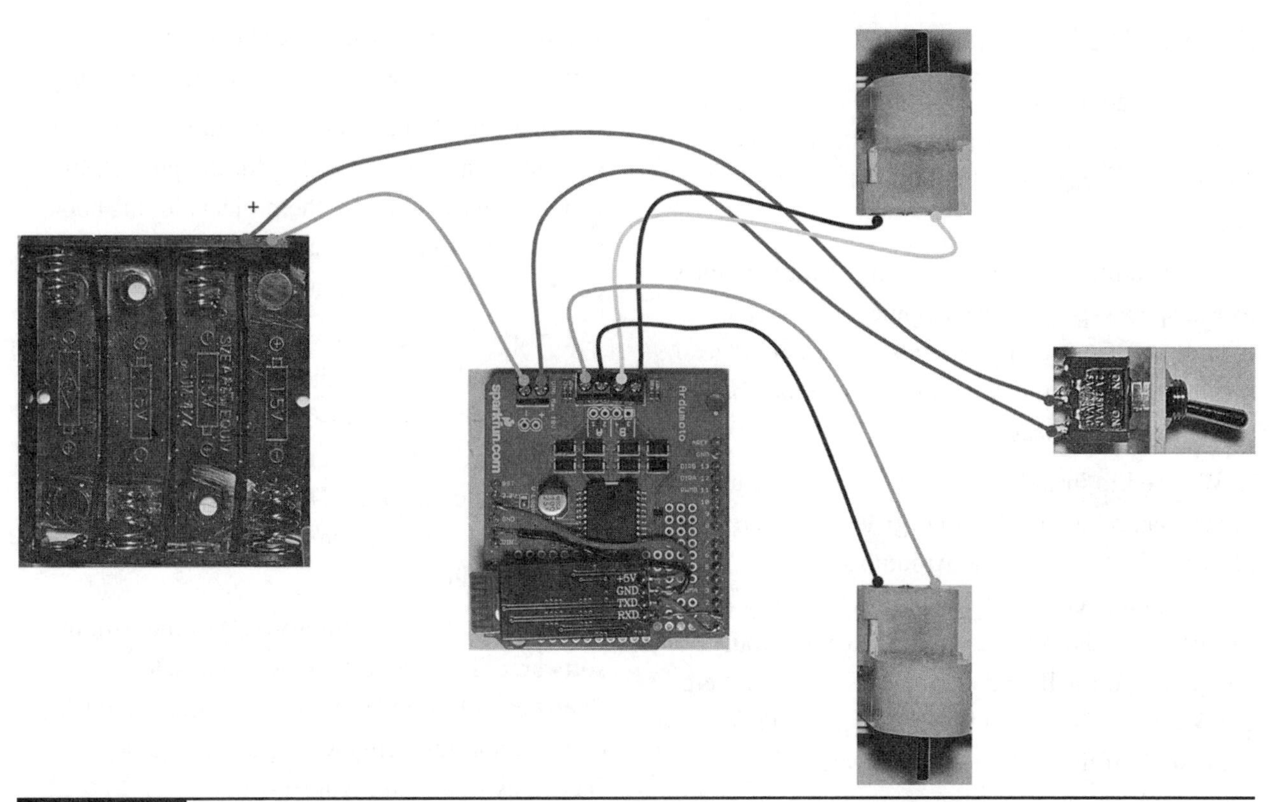

Figure 1-10 The wiring diagram

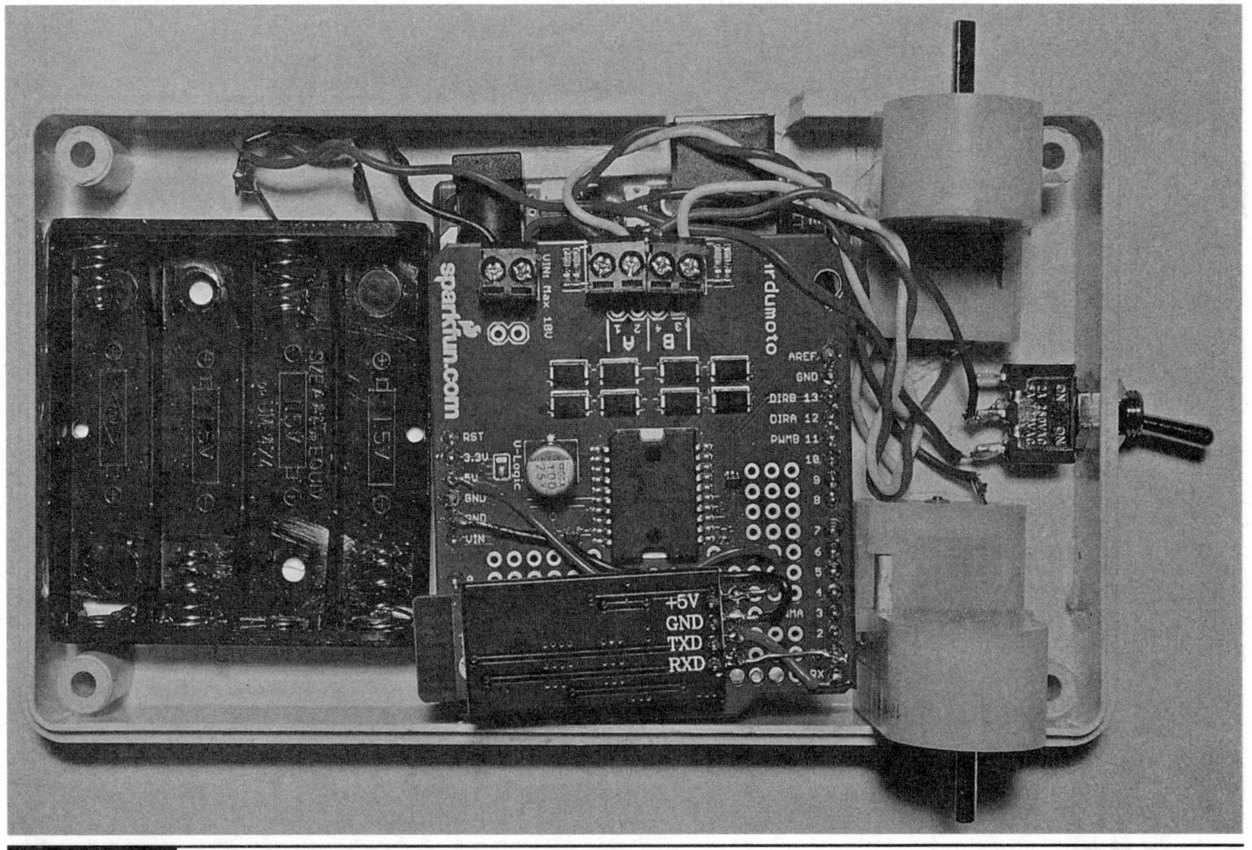

Figure 1-11 Inside the robot

Step 7. Test the Motors

We now need to set up our Arduino environment so we can install a program to test the motors before going ahead and linking it all up with Bluetooth.

The Arduino board we are using (Arduino Uno) uses a special-purpose development environment that allows us to send programs, or "sketches" as they are called in the Arduino world, to the board through the USB lead.

We need to install the Arduino environment, and rather than repeat instructions given elsewhere, please refer to the official Arduino site (www.arduino.cc) and follow the instructions there for installing the Arduino environment on your computer. You will find separate instructions there for Windows, Linux, and Mac. This book uses version 22 of the Arduino software and the Arduino Uno interface board; however, you should have no problem using later versions of Arduino.

Once your Arduino environment is set up, you need to install the test sketch for the project. In fact, all the sketches for the projects in this book are available in a single zip file that can be downloaded from www.duinodroid.com.

Unzip the file and move the whole Arduino Android folder to your sketches folder. In Windows, your sketches folder will be in My Documents/Arduino. On the Mac, you will find it in your home directory, Documents/Arduino/, and on Linux it will be in the Sketchbook directory of your home directory.

After installing the library, restart the Arduino software. Then, from the File menu, select Sketches, followed by Arduino Android, and then ch01_motor_test. This will open the motor test sketch, as shown in Figure 1-12.

Figure 1-12 The motor test sketch

Before we actually run the motors, we may need to change the setting at the top of the script called motorVolts. Set this value to the maximum voltage for your gear motors, if it is different from the Pololu motors—which are nominally 4.5V, but are fine at 5V—you will need to change this value.

The Bluetooth module utilizes the Arduino Rx and Tx pins used by the interface, thus we cannot program the Arduino with the shield connected. So take the shield off for now.

Connect your Arduino board to your computer via USB. We need to tell the Arduino software what type of board we are using, so to set the board, go to the Tools menu and select the Board option. This will give you a list akin to that shown in Figure 1-13.

Select the option for the type of board you are using (Arduino Uno). We then need to do a similar thing for the "Serial Port," which is also part of the Tools menu. This will generally be the top option on the list of ports (COM4 on Windows).

We are now ready to upload the sketch to the board by clicking the upload icon (second from the right on the toolbar). If you get an error message, check the type of board you are using and the connection.

Now that we have programmed the Arduino with the motor test script, detach the USB cable and reattach the shield. Turn on the switch. The motors should now go through the test sequence.

- Both motors forward
- Both motors backward
- Rotate clockwise
- Rotate counterclockwise
- Pause for five seconds

If one of the motors is not working, check the wiring. If you find that one of the motors goes

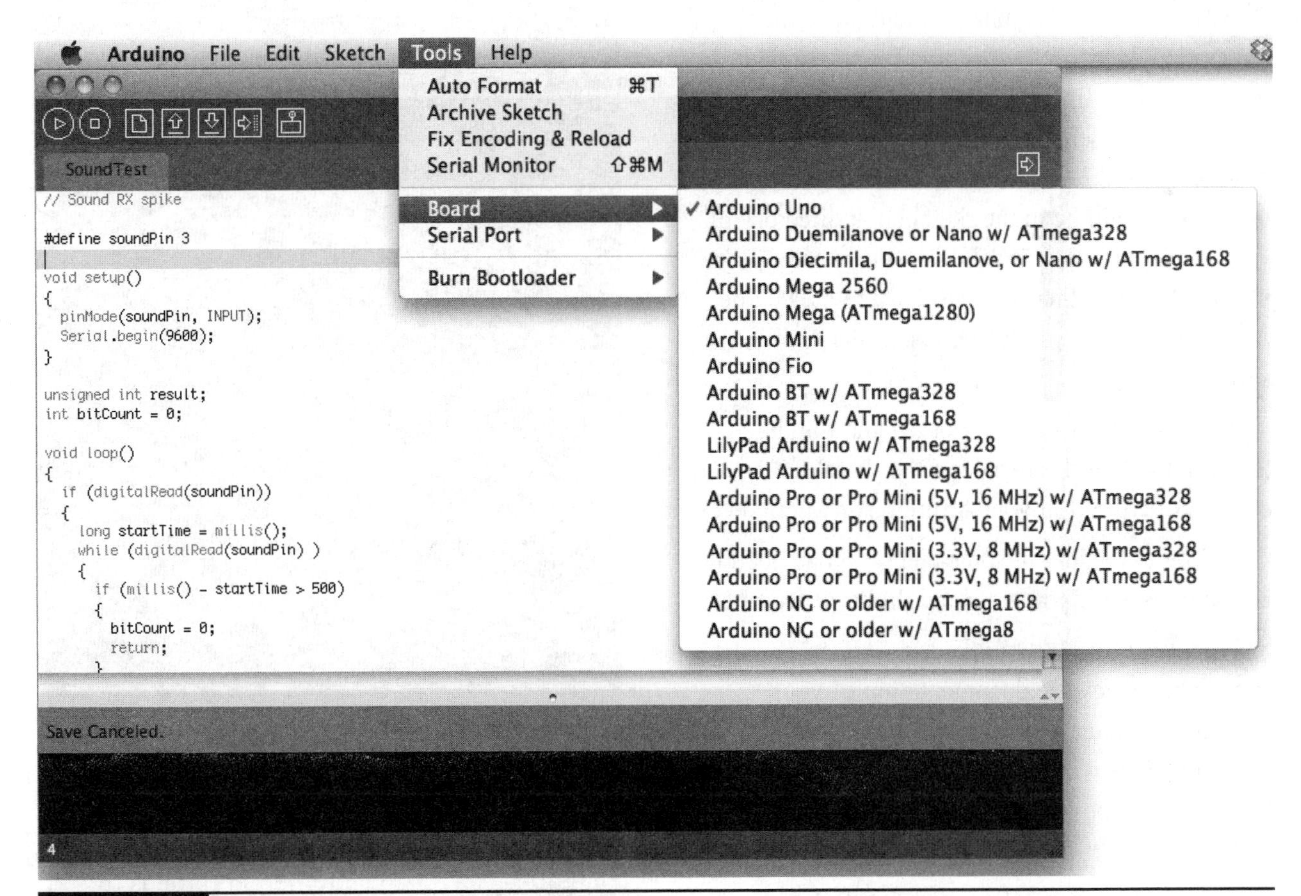

Figure 1-13 Selecting the Arduino board type

forward when it should go backward, swap over the leads at the screw terminals for that motor.

Step 8. Install the Real Arduino Sketch

So far, so good. We can now move on to the next step of installing the real sketch for the motors, which will get its commands from Bluetooth.

The Android app uses a technology called Amarino (www.amarino-toolkit.net). This open-source technology greatly simplifies the writing of Bluetooth Arduino applications. It has two parts, a library that must be installed in your Arduino environment on your computer and an app for the Android phone.

To install the library, go to the downloads page on the Amarino web site (www.amarino-toolkit.net/index.php/download.html) and then click the link for "MeetAndroid - Arduino Library." Download the zip file, unzip it, and move the unzipped folder to your Arduino libraries folder. In Windows, your libraries folder will be in My Documents/Arduino. On the Mac, you will find it in your home directory, Documents/Arduino/, and on Linux, it will be in the sketchbook directory of your home directory. If there is no Libraries folder in your Arduino, then you will have to create one. After installing the software, restart the Arduino software.

Turn the power to the robot off and unplug the Arduino board. Then open the sketch ch01_droid_droid in your Arduino software.

Before uploading it, some changes may need to be made. First, if you are using different motors, change the motorVolts value.

Second, check the documentation of your Bluetooth module to see at which speed it communicates with the Arduino. This is often 9600, but can be faster for some modules.

Finally, you can upload the sketch to the board the same way you did the test script. If you get compilation errors, it is almost certainly due to the Amarino library folder being in the wrong location.

Disconnect the Arduino board from the USB lead and reattach it to the shield. Now we get to the exciting bit!

Step 9. Install the Android App

Unlike the iPhone, you can download your Android applications from anywhere you like. This does mean you have to make sure you are not downloading anything malicious, so you may need to change a setting on your Android device to accomplish this.

Open the Android "Settings" app, navigate to Applications, and check the Unknown Sources box, as shown in Figure 1-14.

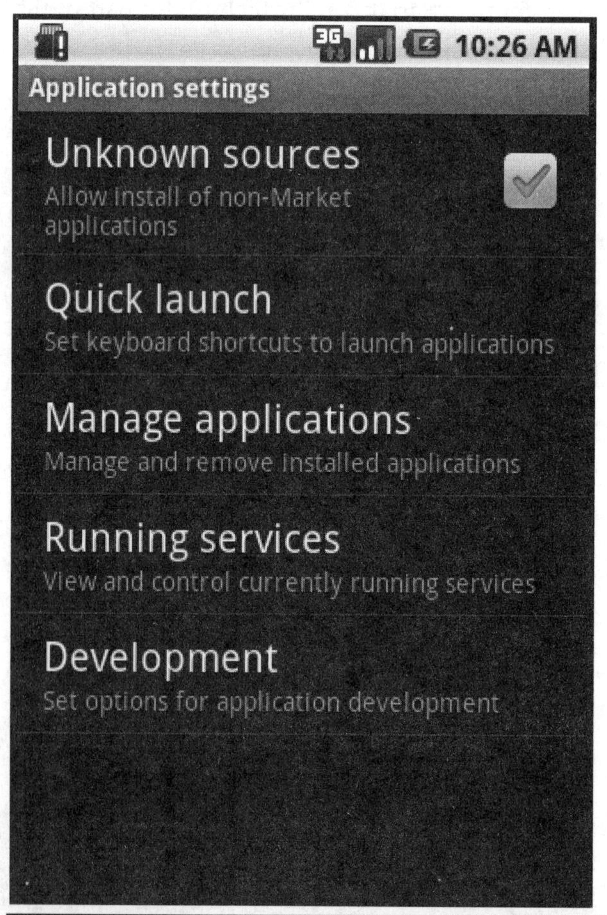

Figure 1-14 Changing Android settings to allow download

To use the app for the robot, we must first install the general-purpose Amarino app, which allows us to manage our Bluetooth devices. This can be downloaded from the Amarino web site, by using your web browser on your Android device and navigating to www.amarino-toolkit.net/index.php/download.html. Then, click the link for "Amarino - Android Application."

To install the robot control app itself, open the browser app on your Android device and navigate to www.duinodroid.com. Click the Downloads tab and then the link for the DroidDroid app.

Step 10. Try It Out!

Before you run the DroidDroid app, we need to run the Amarino app (Figure 1-15).

Figure 1-15 The Amarino app

Power up the robot. You should find that the LED on the Bluetooth module is blinking. This indicates that the module is not yet paired to anything. The Amarino app will allow us to pair it with your phone.

From the main menu shown in Figure 1-15, click the big green Add button. You will then be presented with a list of Bluetooth devices in range.

Select your device from the list. This will return you to the front page with your device added to the list of devices. Clicking Connect will start the pairing process. You will be prompted for a pairing key for the Bluetooth module. This will probably be "1234", but may be "1111". Consult the documentation for the Bluetooth module. Once the pairing is successful, the blinking light on the Bluetooth module should remain lit. Write down the Bluetooth ID for the device. This is the six-part number with colons between each two digits and you will need it in a moment.

Launch the DroidDroid app (Figure 1-16). Next, enter the Bluetooth ID for the device you just

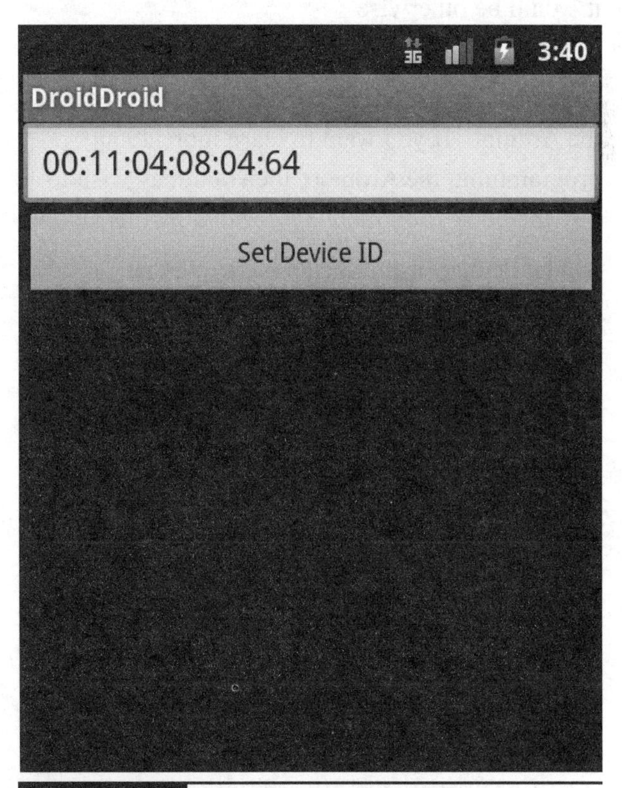

Figure 1-16 Setting the Device ID

noted down from the Amarino app. When you click "Set Device ID," it will launch the main controls (as shown back in Figure 1-1).

Sliding the controls up and down should drive the left and right motors of your robot.

Theory

The software for this and all the other projects in this book are provided as open source and you are encouraged to make your own improvements. The author would very much like to hear about any improvements you make to the software. You can contact the author at www.duinodroid.com.

In this section, we are going to have a quick look at the software used in this project, starting with the Arduino sketch.

The Arduino Sketch

The well-crafted Amarino software makes writing the software for this project a great deal easier than it would be otherwise.

The following description assumes you have an understanding of the C language used to program the Arduino. If you want to learn more about programming the Arduino, then you may wish to buy the book *Programming Arduino* by this author.

The listing for the sketch is given next:

```
#include <MeetAndroid.h>
#define supplyVolts 6
#define motorVolts 5
#define baudRate 9600
MeetAndroid phone;
int left = 255; // midpoint
int right = 255;
int pwmLeftPin = 3;
int pwmRightPin = 11;
int directionLeftPin = 12;
int directionRightPin = 13;
void setup()
{
  pinMode(pwmLeftPin, OUTPUT);
  pinMode(pwmRightPin, OUTPUT);
  pinMode(directionLeftPin, OUTPUT);
  pinMode(directionRightPin, OUTPUT);
  setMotors();

  // use the baud rate your bluetooth
  // module is configured to
  Serial.begin(baudRate);
  phone.registerFunction(setLeft, 'l');
  phone.registerFunction(setRight, 'r');
}
void loop()
{
  phone.receive();
}
void setLeft(byte ignore, byte count)
{
  int value = phone.getInt();
  left = value;
  setMotors();
}
void setRight(byte ignore, byte count)
{
  int value = phone.getInt();
  right = value;
  setMotors();
}
void setMotors()
{
   int vLeft = abs(left - 255) *
     motorVolts / supplyVolts;
   int vRight = abs(right - 255) *
     motorVolts / supplyVolts;
   int dLeft = (left > 255);
   int dRight = (right > 255);
   if (vLeft < 50)
   {
     vLeft = 0;
   }
   if (vRight < 50)
   {
     vRight = 0;
   }
   analogWrite(pwmLeftPin, vLeft);
   analogWrite(pwmRightPin, vRight);
   digitalWrite(directionLeftPin,
     dLeft);
   digitalWrite(directionRightPin,
     dRight);
}
```

The sketch starts with three constants. The values of supplyVolts and motorVolts are used to scale the power supplied to the motors. So if you adapt the design to use different motors of a different battery pack, you will need to change these values.

The baudRate variable should match the baud rate used by the Bluetooth module to communicate with the Arduino.

The interface to the phone is all contained in the MeetAndroid library. To gain access to it, you must create an instance of it—in this case, called "phone".

The variables "left" and "right" are used to hold the speed of each motor. They have a center value of 255. Meaning that at 255 the motor is stopped, at 511 it is full-speed forward, and at 0 it is full-speed reverse.

The next four variables are define the pins used for the motors. These are set by the motor shield, so they cannot be changed. Each motor is controlled by two pins. The "pwm" pin controls the speed of the motor: 0 being stopped; 255 being full speed. The "direction" pin changes the direction: 1 for forward, and 0 for reverse.

The setup function sets the appropriate pin modes and starts the serial port running. It also defines two callback functions—setLeft and setRight—that will be called whenever the sketch receives a message from the Android phone to set a new motor speed with the command letter "l" or "r" for the left and right motors, respectively.

All we actually need in the "loop" function is to call the "receive" function in the MeetAndroid library. This checks for any incoming messages and calls the appropriate callback function.

The two callback functions are responsible for setting a new value for the "left" and "right" variables. The parameters to the callback functions can both be ignored. To retrieve the value sent by the phone, the callback function uses the getInt function.

The setMotors function calculates the appropriate analog output values and scales them to account for the difference between the supply voltage and the motor voltage. It also calculates the direction for each motor and sets the appropriate output values.

The Android App

The Android app is the more complex part of the project, and learning Android programming is a book in its own right. In fact, many such books are available. But at least we can have a little look at the section of code that sends the value to the Arduino.

```
private void updateLeft(){
  Amarino.sendDataToArduino(this,
   DroidDroid.DEVICE_ADDRESS, 'l',
   (511 - left));
}
```

The interface is beautifully simple. You just call the method sendDataToArduino. The first argument is the Android Activity instance (think screen); the second argument is the Bluetooth ID of the device. The next parameter is a single-character flag that will be either "l" or "r", and it is this flag that is used to trigger the appropriate callback function on the Arduino.

The final argument is the integer value sent to the Arduino, which will be a value between 0 and 511.

Summary

This is the first of a series of fun things to do with your phone. In the next chapter, we are going to use Google's new ADK technology to create a Geiger counter accessory for our Android phone. Click-click-click, RUN!!!!

CHAPTER 2

Android Geiger Counter

THE EVIL GENIUS'S MINIONS are always leaving radioactive material lying around after his failed attempts at world domination. Tired of continual accidental exposure to radiation, the Evil Genius decided to make a Geiger counter accessory for his Android phone (Figure 2-1).

Google, developers of the extremely successful Android mobile operating system, chose Arduino as the basis for their open accessory development kit (sometimes called ADK). This is a protocol specification and library software developed by Google to encourage third parties (that's us) to develop hardware accessories to which an Android device can be docked, via its USB connection.

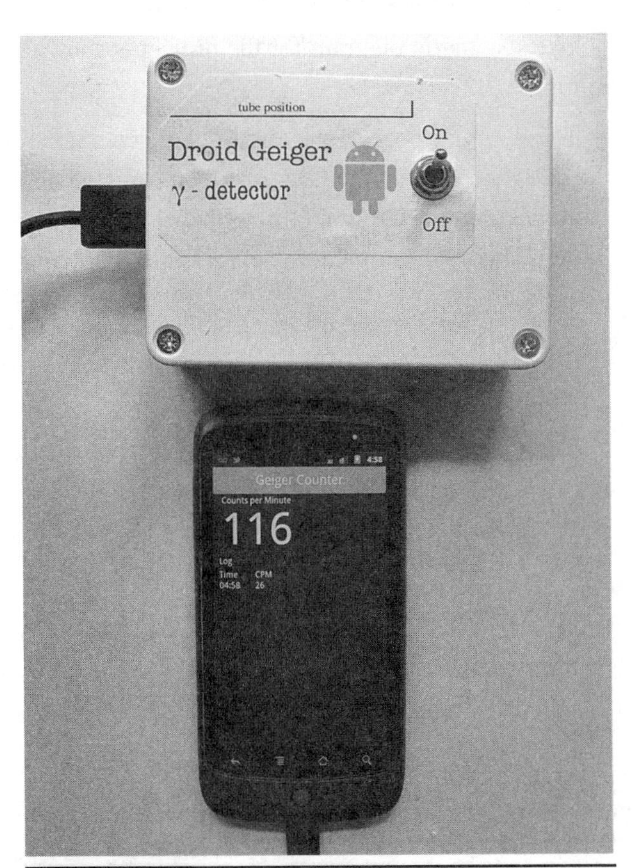

Figure 2-1 The Android Geiger counter

While Google probably expected accessories to be boom boxes and other fairly dull living room–type accessories, the Evil Genius seized the opportunity to develop some more exciting open accessories for his cell phone, such as this Geiger counter.

The Geiger counter uses a cheap Geiger-Müller (GM) tube sourced on eBay for about USD 20. The tube is not sensitive to alpha radiation. Such tubes are more expensive and difficult to obtain, but should still work fine with this design.

The whole project should cost less than USD 100, including the Arduino and USB host shield.

CAUTION **This project generates 400V for the Geiger tube. This voltage will be stored in the capacitors after power has been removed from the device. This can harm you, so take great care when constructing this project. In addition, if you wire this project incorrectly, you could pass a high voltage into your phone and damage it. The author and publisher accept no liability for any damage that may occur to your phone when using this project. Further, do not expose yourself to radiation. Be content to measure the background count.**

Google Open Accessory

The Google Open Accessory Development Kit is based on Arduino technology. You can buy a special development board based on the Arduino, but with LEDs and various other hardware already soldered on the board, you can achieve greater flexibility by using a standard Arduino board and a USB host shield.

Open accessory support is only available in phones that have Android 2.3.4 or later. So before ordering any components, make sure your phone supports Accessory Mode and either has 2.3.4 or can be upgraded to that version or later.

Figure 2-2 shows how things work with the Android and Arduino using the Android ADK.

The Android phone acts as a USB client. That is, it is the Arduino that is in charge of the situation as the USB host. It must initiate the connection with the Android phone. When it does so, it can also trigger the phone to automatically switch into accessory mode and open a custom app. In this case, it will open the DroidGeiger app we have developed for this project.

The Arduino is also required to provide charging power to the phone. So any accessory we build using our Arduino must be capable of providing up to 500mA at 5V over USB to charge the phone. This means that the accessory either needs to be powered from a power adapter, or in the case of this project, powered from a decent battery. Later in this chapter, we also look at how to get around this problem and use a noncharging lead for the connection.

Construction

Like the robot of the previous chapter, this project uses a shield attached to the top of the Arduino Uno. In this case, the shield is a USB host shield. You may be thinking that this is unnecessary because the Arduino has a USB connector. Unfortunately, the USB connection on the Arduino itself is a USB client connection, and to connect to an Android phone we need a USB host connection. In actual fact, the USB host shield has a handy prototyping area to which we can solder the other components needed for the project.

The schematic diagram for the project is shown in Figure 2-3.

The main purpose of the circuit is to generate the 400V needed by the GM tube. You can find a more detailed description of how this circuit works in the "Theory" section at the end of this chapter.

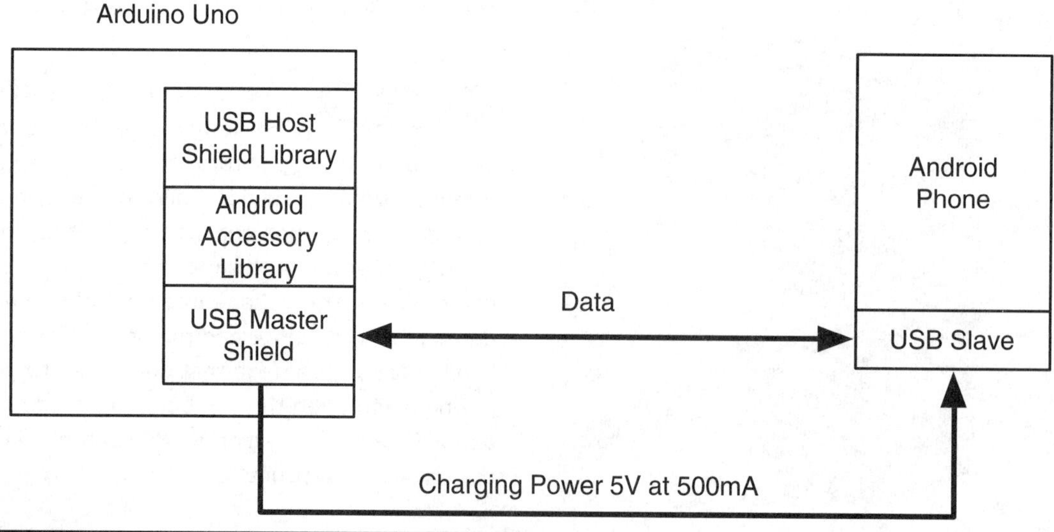

Figure 2-2 Arduino and Android, together at last

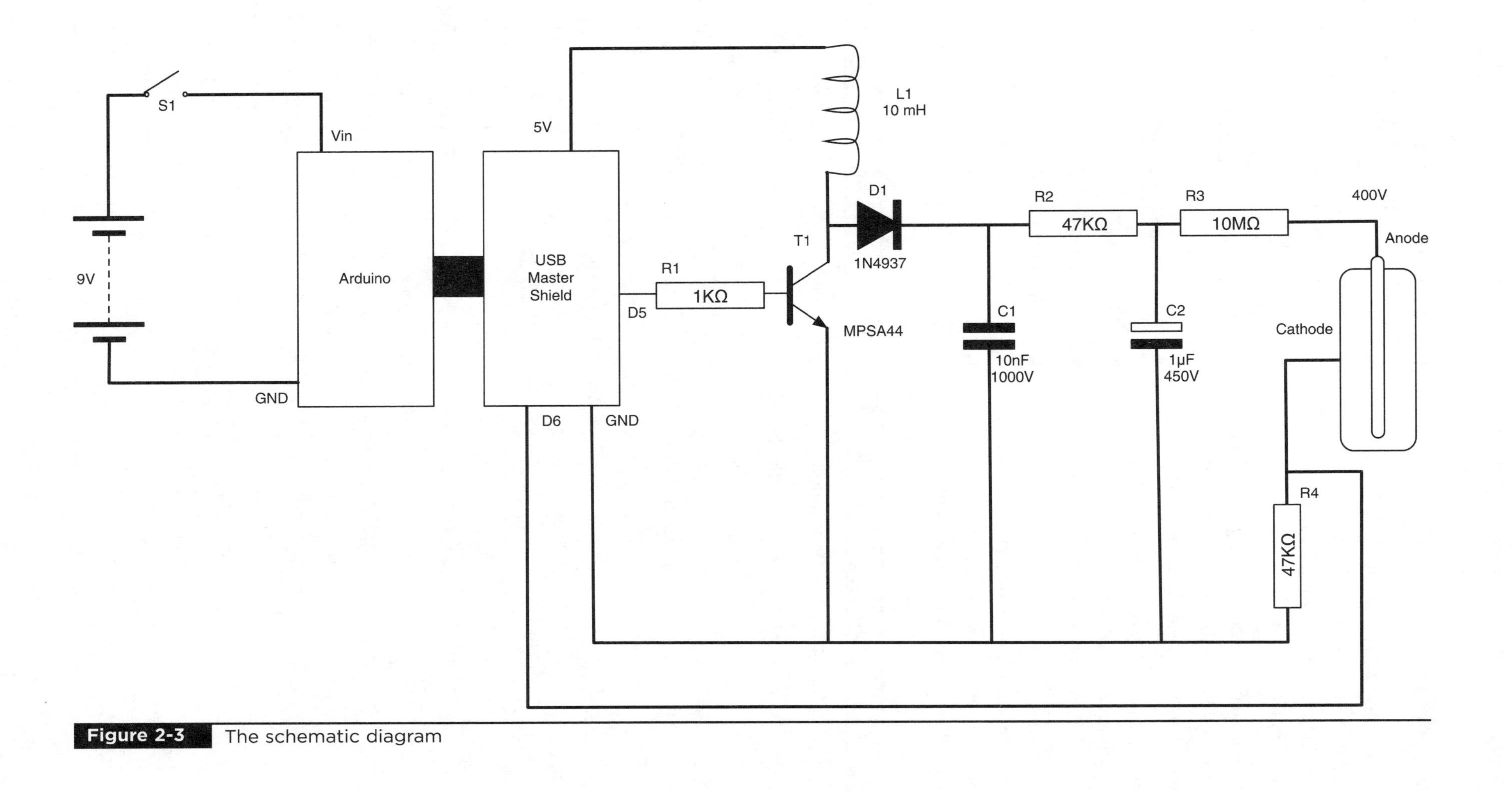

Figure 2-3 The schematic diagram

What You Will Need

In addition to an accessory-capable Android phone (Android 2.3.4 or later), you will need the components listed in the following Parts Bin to make the sound link module.

If you plan to make the noncharging USB lead, you will need two 1kΩ resistors rather than just one. You will also need an old USB extension lead.

For a longer-lasting battery alternative, use a holder that will accept six AA cells to supply the project with 9V. This alternative is shown in some of the figures.

The key component for this project is the GM tube. These can be easily found on eBay from international sellers, often from former Soviet Union countries. The tube that the author described is a "Russian Military GEIGER TUBE COUNTER CI-1."

The tube has the following specifications. However, other similar devices should work just fine. Our Geiger counter is not for critical situations. It will not be calibrated, so accuracy is not really a consideration.

- Type: Gamma detector
- 360–440V anode voltage
- Plateau length: 80V
- Inclination: 0.125%/V
- Load resistance: 10MΩ

PARTS BIN

Part	Quantity	Description	Source
Arduino Uno	1	Arduino Uno board	www.arduino.cc
USB shield	1	Arduino USB Host Shield	Sparkfun: DEV-09947
Pin headers	1	Pin header strip; broken into two sections of six pins, and two sections of eight pins	Farnell: 1097954
Switch	1	SPST miniature toggle switch	Farnell: 1661841
Battery	1	PP3 9V NiMH rechargeable battery	
Battery clip	1	PP3-style battery clip	Farnell: 1183124
Case	1	Plastic case	
R1 plus 1	2	1kΩ 0.5W metal film resistor	Farnell: 9339779
R2, R4	2	47kΩ 0.5W metal film resistor	Farnell: 9340637
R3	1	10MΩ 0.5W metal film resistor	Farnell: 1779379
C1	1	10nF 1000V ceramic capacitor	Farnell: 1615007
C2	1	1μF 450V electrolytic capacitor	Farnell: 1822752
D1	1	1N4937	Farnell: 9843663
T1	1	MPSA44	Farnell: 1574391
L1	1	10mH choke	Farnell: 1710435
Tube	1	GM tube (see above)	eBay
Clips	2	Fuse clips	Farnell: 1866097
Power plug	1	2.1mm power plug	Farnell: 1200147

- Length: 90mm
- Diameter: 12mm

We will see later how to adjust the anode voltage we generate using the Arduino sketch between 0 and 450V, the limit for our capacitors.

This design employs the Arduino Uno. The official Arduino web site (www.arduino.cc) lists suppliers of the Uno. However, if you are on a budget, you can use an Arduino clone of the Arduino Uno. The Arduino is open-source hardware, which means that all the design files are available under a Creative Commons license that permits third parties to make their own Arduinos. Many do, and an Internet search will reveal cheap alternatives to the official Uno.

In addition to these components, you will also need the tools listed in the following Toolbox.

TOOLBOX

- An electric drill and assorted drill bits
- A hacksaw or Dremel rotary tool
- A hot glue gun or epoxy glue
- Assorted self-tapping screws
- A computer to program the Arduino
- A USB-type A-to-B lead
- A multimeter with a 1000V range

Step 1. Attach Pin Headers to the Shield

The first step is to attach the pin headers to the motor shield. Figure 2-4 shows the shield with the pin headers. Your header strips will probably come in a single long length that is designed to be

Figure 2-4 The USB shield with pin headers

broken into sections of the correct length. You will need to break off two lengths of six pins and two lengths of eight pins.

The easiest way to keep the pins straight is to plug the long ends of the headers into your Arduino board while you solder them. However, this will heat up the plastic of the socket underneath, which may distort it. So either solder quickly, or just solder the pins at each end of a section so the header is held in the right place, and then remove the shield and solder the rest of the pins.

When all the pins are in place, the top of the shield should look like Figure 2-5.

Figure 2-5 The top side of the USB shield attached to an Arduino

Step 2. Attach the Low-Lying Components

Figure 2-6 shows the layout of the components on the prototyping area of the shield.

It is always easiest to solder the lowest-lying components first, so we start by soldering the resistors and diode into place. Do not cut off the excess lead on the underside of the board, as we are going to use these to connect up the components.

The diode must be the correct way around, which is with the stripe towards the USB socket on the shield.

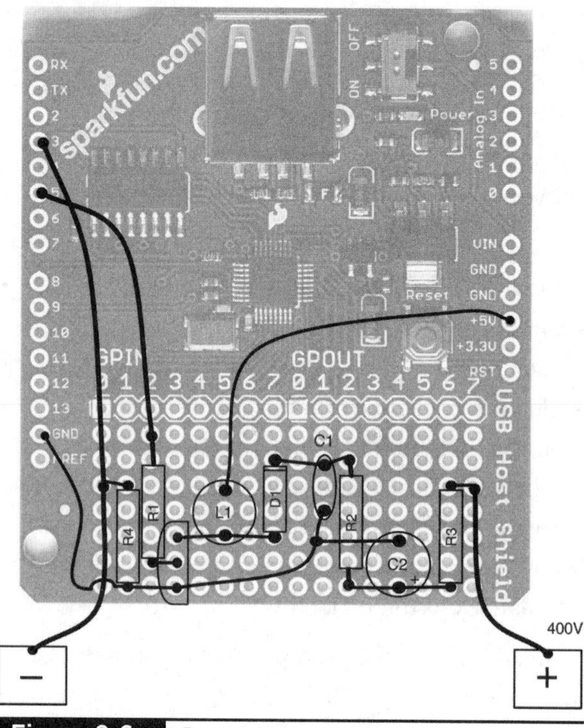

Figure 2-6 Prototype area layout

Figure 2-7 shows the resistors and diode in place.

Figure 2-7 Resistors and diode on the shield

Step 3. Solder the Remaining Components

We can now solder the remaining components (Figure 2-8).

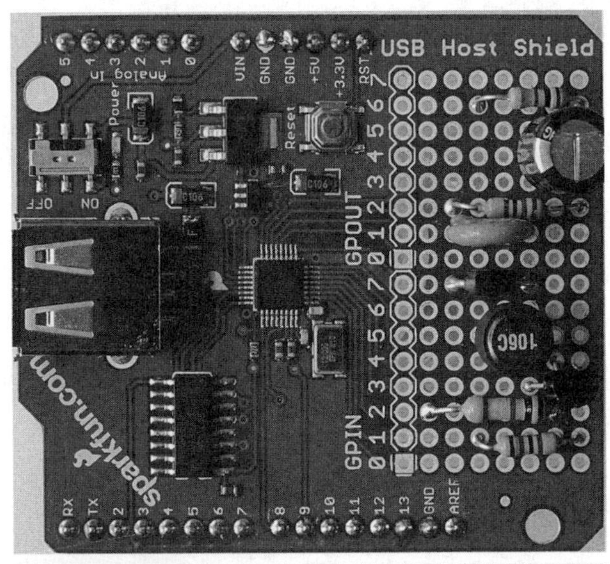

Figure 2-8 The remaining components on the shield

Take special care to get the transistor the correct way around. The electrolytic capacitor C2 must also be the correct way around. The longer positive lead should be directed toward the outside edge of the board. The inductor can be placed either way around.

Once all the components are in place, the bottom of the board should look like Figure 2-9. The next stage is to bend and shorten the leads of the components to make all the connections between the components shown previously in Figure 2-6.

Figure 2-9 The bottom of the board with all the components in place

Figure 2-10 The component leads used to make connections

After that, the bottom of the board should look something like Figure 2-10.

Step 4. Solder the Leads to the Arduino Pins

The final connections we need to make involve soldering some leads from the prototyping area to the appropriate Arduino pins and then to the fuse holder sockets to connect the GM tube. Figure 2-11 shows these connections. The leads can go through

Figure 2-11 Clips for the GM tube using fuse holder sockets

holes in the prototype area near the component leads that they need to be connected to. Again, use Figure 2-6 as a reference to see how the leads should be connected.

Do not connect the GM tube yet. We need to make some tests first.

Step 5. The Final Wiring

The project is supplied power from either a 9V PP3 battery or a battery holder containing six AA batteries. Since the project will charge the Arduino while it is on, it makes sense for these to be rechargeable batteries. Either way, the battery box uses a PP3-style clip, so we need to put the switch in line with the positive battery lead as per Figure 2-12.

As you can see from Figure 2-12, the power lead ends in the 2.1mm power plug that will supply power to the Arduino and its shield. Alternatively, if space is tight, you can solder the leads directly to the Vin and GND sockets on the shield, as shown in Figure 2-17.

Step 6. Install the Open Accessory Libraries

If you have not already done so, install the Arduino software on your computer. You can find comprehensive instructions for this on the official Arduino web site at www.arduino.cc.

The Google Open Accessory requires that two libraries be installed into your Arduino environment. The first of these is a version of the USB host library, which is patched to work with standard Arduino hardware. This should be downloaded from microbridge.googlecode.com/files/usb_host_patched.zip.

If for any reason, you cannot find a download for any of the software used in these projects, please refer to the book's web site (www.duinodroid.com) where there will be instructions for obtaining the software elsewhere.

To install the library, download the zip file, unzip it, and move the unzipped folder to your Arduino libraries folder. In Windows, your libraries folder will be in My Documents/Arduino.

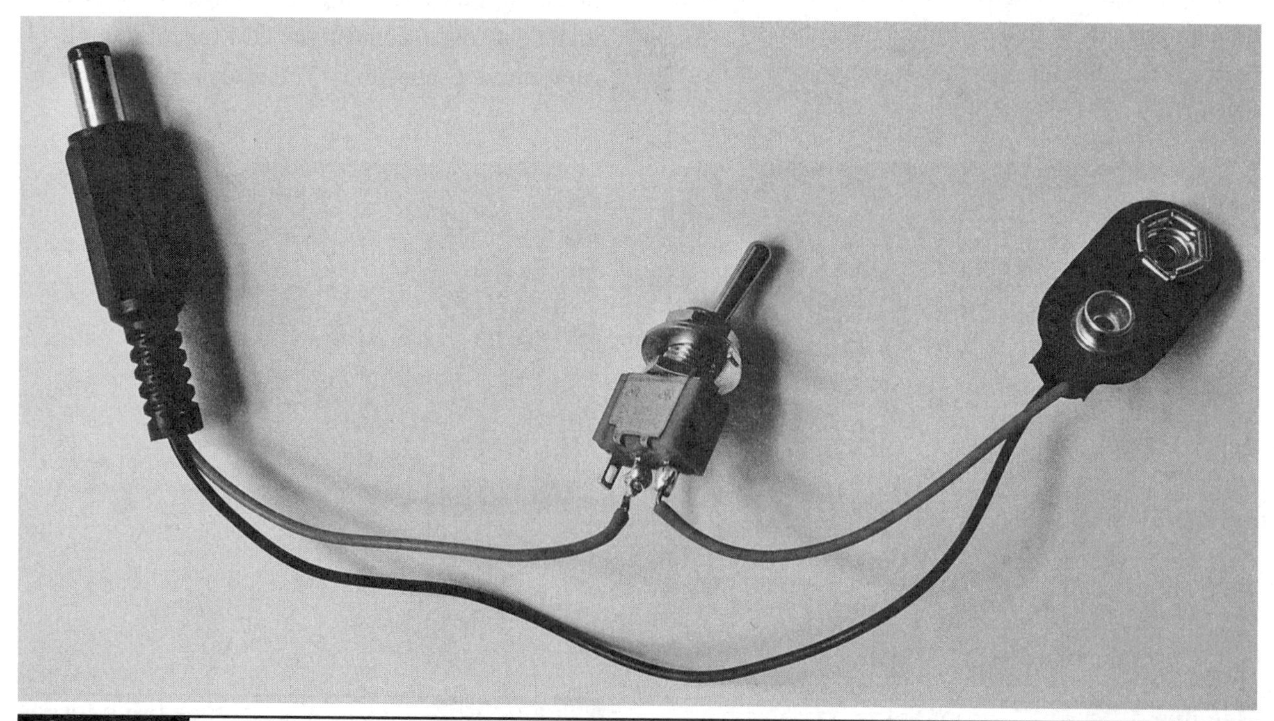

Figure 2-12 The battery lead and power socket

On the Mac, you will find it in your home directory, Documents/Arduino/, and on Linux, it will be in the sketchbook directory of your home directory. If there is no Libraries folder in your Arduino, then you will have to create one. After installing the software, restart the Arduino software.

The second library—the AndroidAccessory library itself—is downloaded as part of the "Adk package download," which can be found at http://developer.android.com/guide/topics/usb/adk.html.

Click the link for "Adk package download." This will download a zip file. Unzip it and you will find a folder inside named "ADK_release_0512." It has a couple of files and three folders. The only folder we are interested in though is the one called "firmware." This contains a folder called "arduino_libs" and within that are two folders, each containing an Arduino library. One is the "USB_Host_Shield," which we do not need to install, as we have already just installed a version of that. However, we do need to install the AndroidAccessory library.

To do this, just move the whole "AndroidAccessory" folder to your Libraries folder the same way you did for the USB host library.

You will need to restart the Arduino software for it to pick up the new libraries.

Step 7. Install the Arduino Sketch

If you built the Bluetooth robot of Chapter 1, you will have already downloaded the zip file containing all the sketches from www.duinodroid.com. If you have not downloaded this, do so now. Unzip the file and move the whole Arduino Android folder to your sketches folder. In Windows, your sketches folder will be in My Documents/Arduino. On the Mac, you will find it in your home directory, Documents/Arduino/, and on Linux, it will be in the sketchbook directory of your home directory.

You will need to restart the Arduino software for it to pick up the new sketches.

From the File menu of the Arduino application, select Sketchbook, then Arduino Android, and then the sketch ch02_Geiger_counter.

Connect your Arduino board (without the shield attached) to your computer via USB. We need to tell the Arduino software what type of board we are using, so, to set the board, go to the Tools menu, and then choose the Board option.

Select the option for the type of board you are using (Arduino Uno). We then need to do a similar thing for the Serial Port option, which is also on the Tools menu. This will generally be the top option on the list of ports (this is COM4 in Windows).

We are now ready to upload the sketch to the board by clicking the "upload" icon (second from the right on the toolbar). If you get an error message, check the type of board you are using and the connection.

Step 8. Test the High-Voltage Supply

The reason we have not attached the GM tube yet is because we need to make sure we are generating the correct high voltage for it, and then tweak it as necessary.

Disconnect the Arduino from your computer and plug the shield into the Arduino board. Note that from now on, whenever there is power supplied to the Arduino, either through the USB connector or the battery pack, parts of the shield will be at high voltage, so be careful not to touch any bare wires. Not only that, but the circuit will retain a charge and stay at a high voltage for some considerable time after the power has been turned off.

Set your multimeter to its 1000V DC range and attach the test leads to the test points shown in Figure 2-13. The negative lead should go to the bottom of R4 (GND) and the positive lead to the bottom of R3 (not the side connected to the positive clip for the GM tube).

If you have test probes rather than clips, then you will have to touch them to the test points once the power is on.

Plug the battery into the Arduino and shield combination and turn on the switch. If all is well, you should see a voltage of around 400V. If the voltage is within the tolerance of your GM tube, you are done.

If the voltage is too high or too low, you will need to adjust the sketch to change the voltage.

Look at the top of the sketch. You will see the line:

```
int op = 45;
```

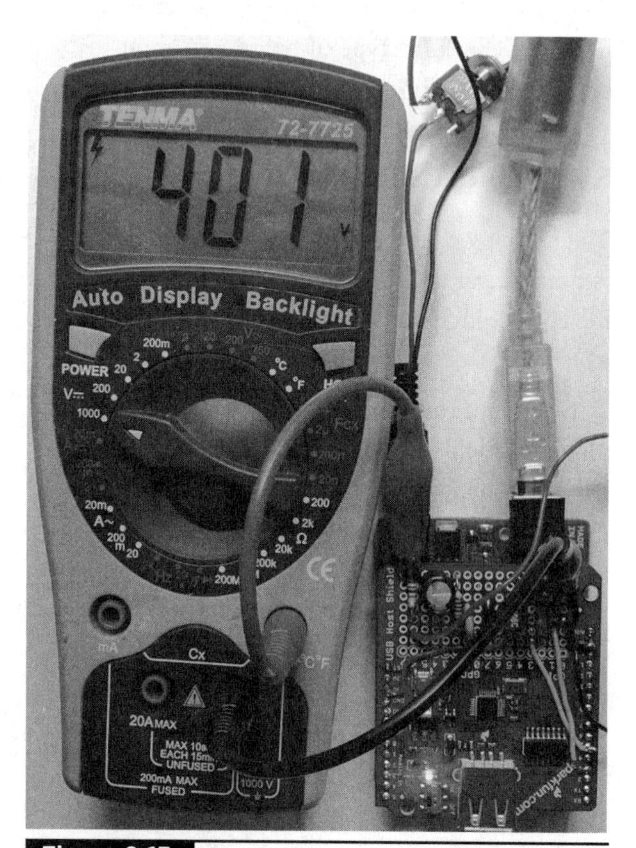

Figure 2-13 Testing the high-voltage output

We can adjust this value up or down (do not increase it above 200) to change the voltage up or down, respectively.

The following table shows how the voltage changed using the author's components. You will probably find your results to be a little different.

Value of "op"	Voltage (V)
45	400
50	425
70	500

Finding the right value is a matter of trial and error. You could carefully plug in the USB lead to your computer and upload the sketch again with a changed value of "op" without turning everything off. If you do this and a freak misconnection puts 400V though your USB port and destroys your computer, then don't say you weren't warned. The safer alternative is to unplug everything, and using insulated pliers, hold the two connectors for the tube together for half a minute to discharge the capacitors.

Step 9. Install the GM Tube

Unplug and discharge everything since we are now ready to attach the GM tube and see if it is working. Fit it into the fuse holders, making sure that the positive end of the tube goes to the positive clip. The positive end of the tube is often marked with a red dot or a + sign. If in doubt, refer to the documentation that came with your tube.

Step 10. Install the Android App

Installing the Android app itself is now just a matter of visiting www.duinodroid.com on your phone's browser and following the download link for the DroidGeiger app. Note that you may need to enable the Android option to allow apps from unknown sources as described in Step 10 of Chapter 1.

Step 11. Test

Once the app is installed, we can lay everything out on the bench, as shown in Figure 2-14.

Plug your phone into the Host USB connection of the shield. After a few seconds, this should trigger the DroidGeiger app to automatically start and begin displaying an average reading. Every time there is an event, the phone will make a click noise and the radiation symbol on the right will flash.

The average reading is based on the gap between events, which is then smoothed. The algorithm definitely has room for improvement. Keep an eye on the book's web site for improvements to this sketch by the author or other readers.

Once a minute, a new entry will appear in the Log area at the bottom of the screen, showing the number of events (counts) in the last minute.

Step 12. Box the Project

In selecting a box for this project, it needs to accommodate the battery, the Arduino board, and the shield, and leave enough room for both the switch and the GM tube.

The box will also need holes in it for the switch, for access to the USB host socket, and four holes for attaching the Arduino board using four small self-tapping screws (Figure 2-15 and Figure 2-16).

The GM tube itself can be accommodated inside the enclosure as the gamma radiation that it detects

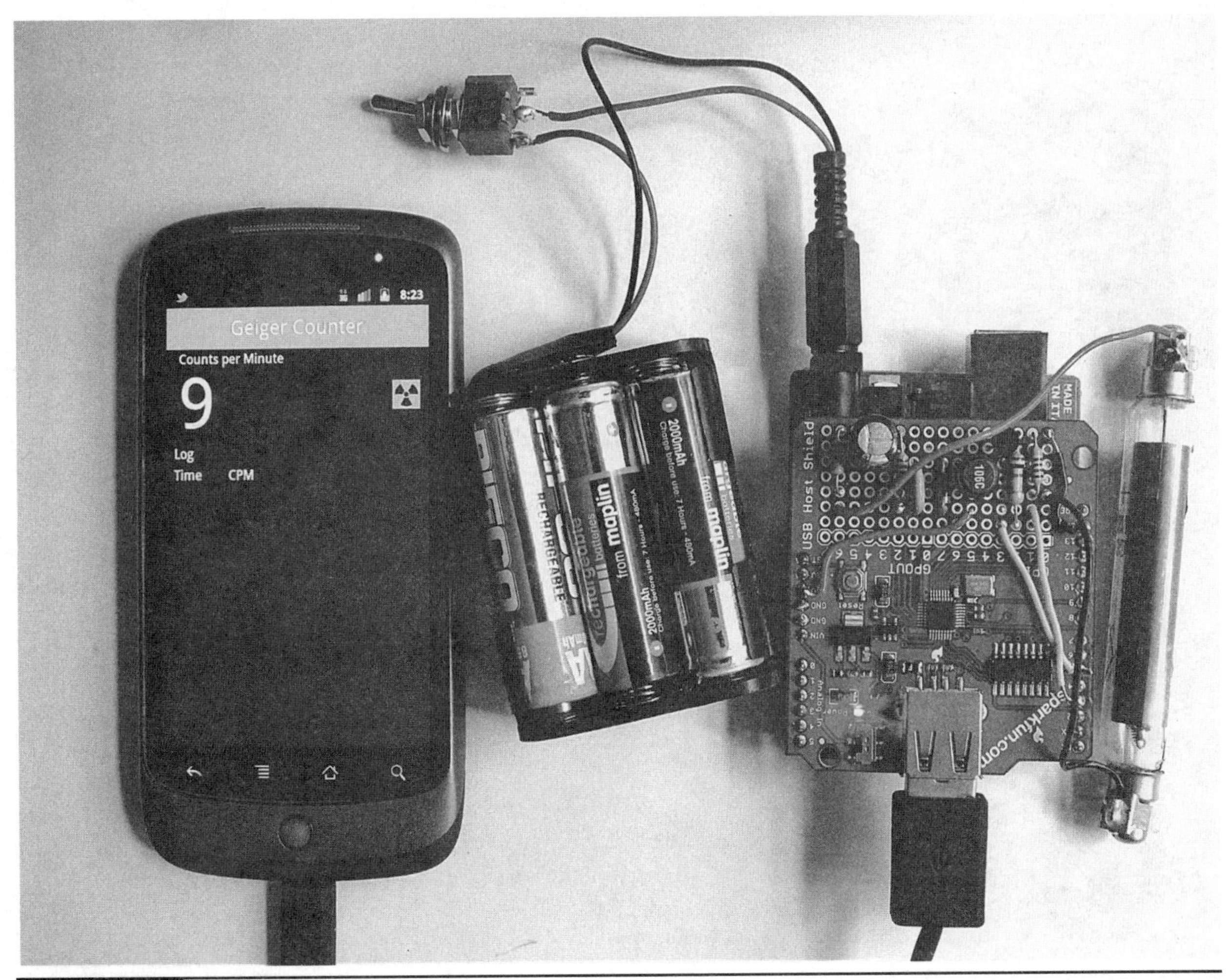

Figure 2-14 Testing the Geiger counter

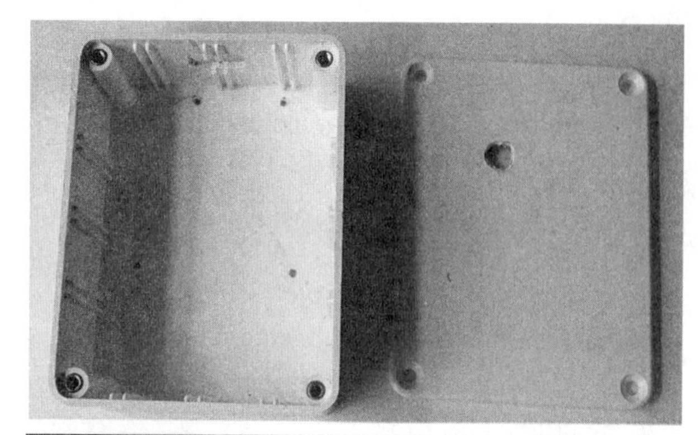

Figure 2-15 The box, drilled

is not much attenuated by a plastic box. To stop it rattling around in the box, a cable-tie can be used to attach it to the USB shield (Figure 2-17).

If you get yourself a tube that is sensitive to alpha radiation, you will need to make a hole for the end of the tube, because Alpha radiation does not penetrate plastic.

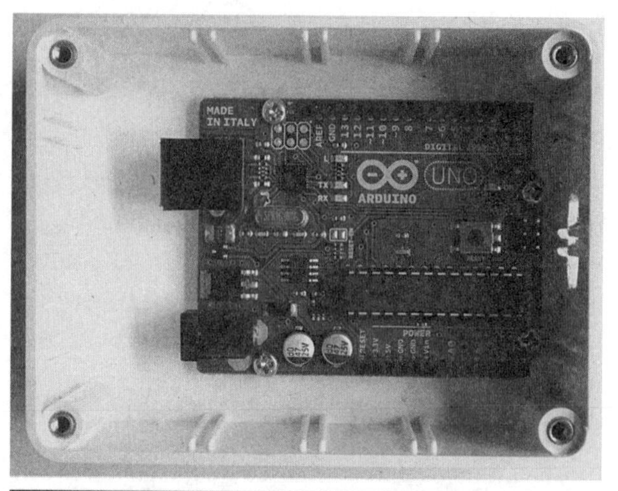

Figure 2-16 The Arduino board fixed into the box

The battery is held in place by the USB socket on the Arduino board. All that remains is, as a final flourish, to print out a paper label for the front of the box (see Figure 2-1). You can find a

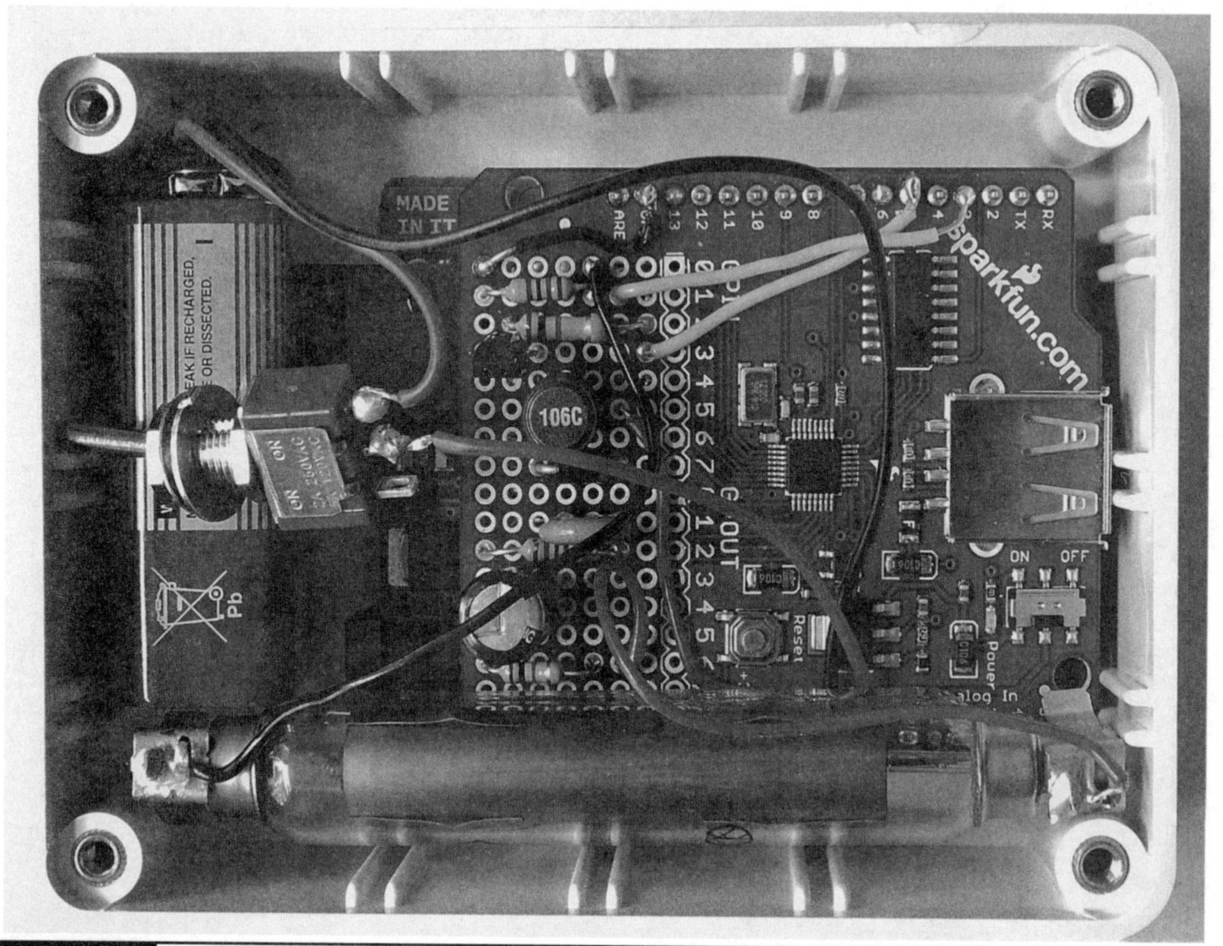

Figure 2-17 The components inside the box

template for this on the book's web site at www.duinodroid.com.

This project uses a small battery. In fact, if you opted to use the small 9V battery, it is really too small to supply the 500mA charging current that the Open Accessory says the accessory should supply. To get around this, either just use the Geiger counter when your phone is fully charged, or construct a "noncharging" lead by splicing a 1kΩ resistor into the +5V supply of an unwanted USB extension cable. This will allow the accessory to work, without the phone being able to draw any significant current from it.

The following sequence of photos shows the construction of the lead.

Carefully cut away some of the outer plastic with a sharp knife on one side and tease out the shielded bundle of wires, scrape a gap between the shielding to get at the +5V wire, and cut the red wire and strip and tin the ends (Figure 2-18).

Solder the 1kΩ resistor into place. Then wrap it all up with insulating tape. Wrap the part behind the tape carefully, so there is no chance of the bare leads from either side of the resistor making contact with the shielding wires or foil. (See Figure 2-19.)

You can then use this lead in between your phone and the accessory, on this and on the rest of the Open Accessory projects in this book.

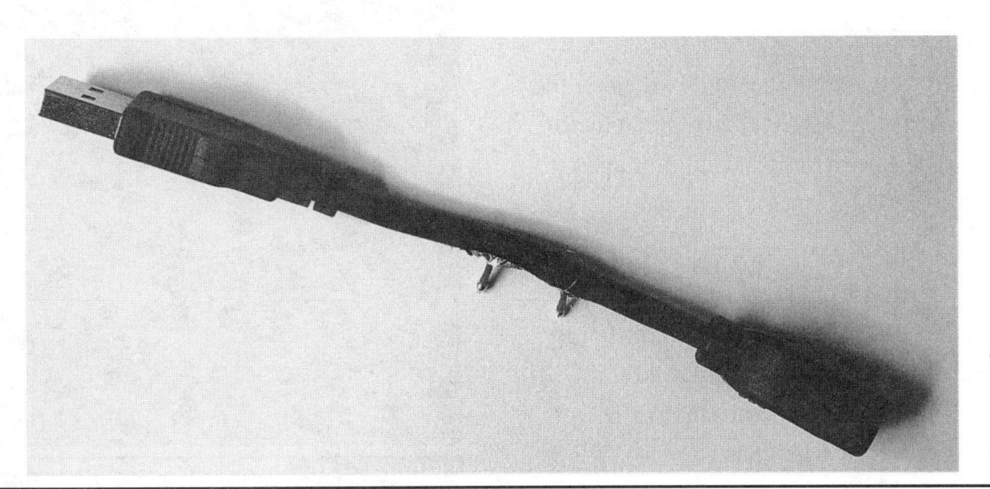

Figure 2-18 Making a USB nonpowered lead

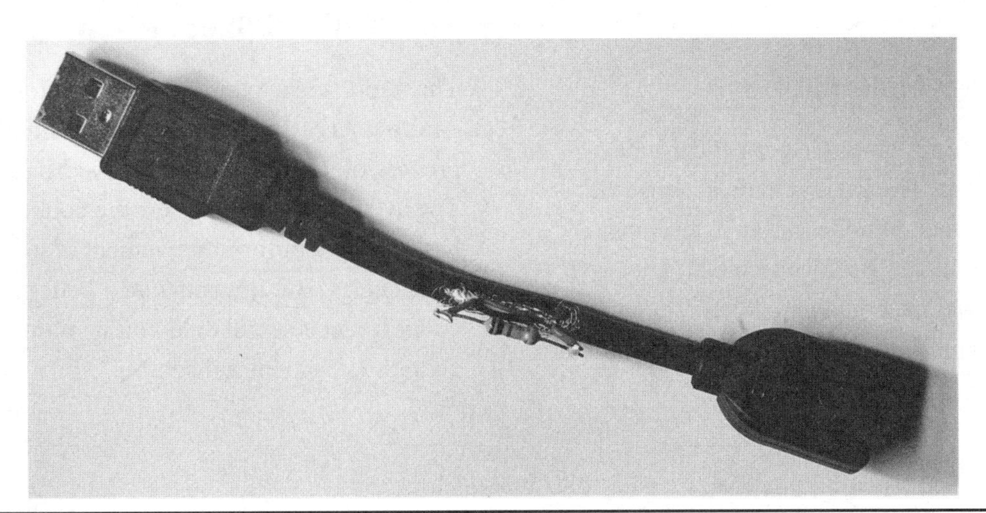

Figure 2-19 Fixing the resistor in place

Theory

The software for this and all the other projects in this book are provided as open source, and you are encouraged to make your own improvements. The author would very much like to hear about any improvements you make to the software. Ways to contact the author can be found at www.duinodroid.com, where you will also find all the source code for both Android and Arduino.

In this section, we are going to have a quick look at the software used in this project, starting with the Arduino sketch. But, before that, we will look at how we generate the 400V needed by the GM tube.

Generating 400V

The circuit used to generate the 400V relies on the Arduino to generate pulses that drive the inductor via a high-voltage transistor. The high-voltage pulses are harvested and smoothed by D1, C1, C2, and R2.

The pulses coming from the Arduino are generated from a PWM output. This kind of digital output from the Arduino is generally used for things like controlling the speed of motors by sending pulses of varying lengths.

We use the same mechanism to tweak the voltage generated.

The Geiger-Müller Tube

For a full description of Geiger-Müller tubes, visit Wikipedia. The basic principal is that the tube has an outer cathode (negative terminal) and an inner anode (positive terminal) enclosed in a glass envelope containing a gas. The more expensive "alpha" detecting tubes have a thin mica end that allows the alpha particles to enter. These are stopped by the glass in the type of tube the author used.

When a particle enters the tube, it ionizes the gas inside, causing a pulse of conduction that can be detected.

You can see these pulses if you connect an oscilloscope across R4. Figure 2-20 shows this.

The timebase of the oscilloscope is set to 500 ms, making the pulses about two seconds apart. The Y scale is set to 2 volts/division, giving the pulses an amplitude of about 5V. The pulses are extremely narrow.

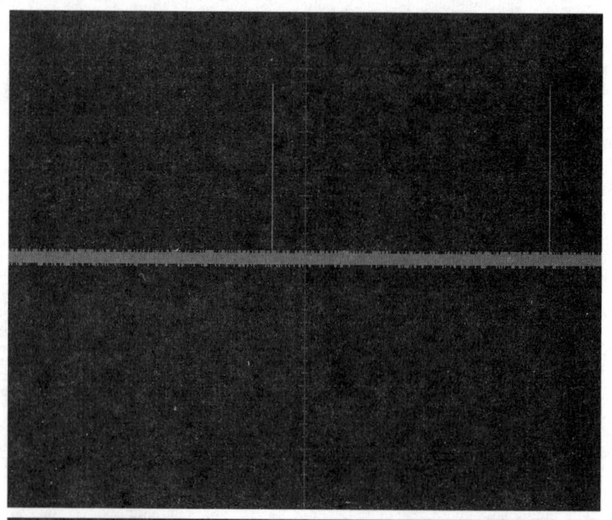

Figure 2-20 An oscilloscope trace of events

The Arduino Sketch

The Arduino sketch is quite straightforward. The Android Accessory library greatly simplifies the process of sending data to the Android device. The averaging algorithm for the counts per minute has much room for improvement at low-count frequencies. But it works much better at high-count frequencies, should you be unfortunate enough to encounter them.

```
#include <Max3421e.h>
#include <Usb.h>
#include <AndroidAccessory.h>

int oscPin = 5;
int op = 45;
int minPulseSep = 50;
long lastEventTime = 0;
long lastTimerTime = 0;
long timerPeriod = 500l;
long lastLogTime = 0;
long logPeriod = 60000l;
int count = 0;

float smoothingFactor = 0.6;
float instantaneousCPM = 0.0;
float smoothedCPM = 0.0;

AndroidAccessory acc("Simon Monk",
     "DroidGeiger",
     "Geiger Counter Accessory",
     "1.0",
     "http://www.duinodroid.com",
     "0000000012345678");

void setup()
{
  Serial.begin(9600);
  pinMode(oscPin, OUTPUT);
  analogWrite(oscPin, op);
  acc.powerOn();
  attachInterrupt(1, eventInterrupt, RISING);
}

void loop()
{
  if (acc.isConnected())
  {
     // every half second, take the instantaneous reading and integrate
     // it with the average reading, then send it
     long timeNow = millis();
     if (timeNow > (lastTimerTime + timerPeriod))
     {
       lastTimerTime = timeNow;
       integrateInstantReadingIntoSmooth();
       sendMessage('R', (int) smoothedCPM);
     }
     // every minute, send the accumulated minute total
     timeNow = millis();
     if (timeNow > (lastLogTime + logPeriod))
```

(continued)

```
        {
            lastLogTime = timeNow;
            sendMessage('L', count);
            count = 0;
        }
    }
    delay(100);
}

void eventInterrupt()
{
    // set instantaneous Reading
    calculateInstantCPM();
    count ++;
    sendMessage('E', 0);
}

void calculateInstantCPM()
{
    // instantaneous cpm = 60,000 / dt in mS
    long timeNow = millis();
    long dt = timeNow - lastEventTime;
    if (dt > minPulseSep)
    {
        instantaneousCPM = ((float)logPeriod) / dt;

       lastEventTime = timeNow;
    }
}

void integrateInstantReadingIntoSmooth()
{
    smoothedCPM = smoothedCPM * smoothingFactor
        + instantaneousCPM * (1 - smoothingFactor);
}

void sendMessage(char flag, int cpm)
{
    if (acc.isConnected())
    {
        byte msg[4];
        msg[0] = 0x04;
        msg[1] = (byte) flag;
        msg[2] = cpm >> 8;
        msg[3] = cpm & 0xff;
        acc.write(msg, 4);
    }
}
```

We start by including the libraries we need. The Max3124e library is included as part of the USB host library you have already installed.

There is then a pin definition, "oscPin," which is for the oscillator signal from the Arduino to drive the coil to generate the high voltage. Note that there is no pin defined for the pulse signal back when an event is detected. That is because it is attached to an interrupt on pin 3.

The op variable controls the pulse width, as described earlier in this chapter.

The "long" variables named "last" something are all used to record times. The function millis() returns a long (32-bit) number representing the number of milliseconds since the Arduino board was reset. This is used in timer code to determine if certain actions are due.

The variables timerPeriod and logPeriod are constants representing half a second and one minute, respectively, and are used in the loop code to carry out the periodic actions of sending a regular update to the Android device (every half second) and sending a count for the minute.

The next three "float" variables are all used in providing an average instantaneous value to send to the phone.

After defining the variables, the line beginning "AndroidAccessory" sets up the connection between the phone and the Arduino. The parameters are:

- Manufacturer
- Model
- Description
- Version
- URI
- Serial number

Of these, only manufacturer, model, and version must match with the corresponding entries in the manifest of the Android project to cause the Android app to launch.

The others may, however, be displayed on the phone.

The "setup" function starts up serial communications. This is not required for the sketch to function; however, the USB library will report the connection-making progress through this, so if you are having trouble connecting to your phone, then opening the Serial Monitor may give you some clues.

A couple of other interesting things are going on in the "setup" function. The "acc.powerOn" tells the phone that the accessory is on and ready. We then attach an interrupt. The first argument is the interrupt number. Interrupt 1 corresponds to pin 3. Whenever there is a rising signal (0 going to 5V) on pin 3, the processor will be interrupted and the function "eventInterrupt" will be run.

The loop function first calls "acc.isConnected" that both tests to see if there is a connection and attempts to establish one if there isn't. If the accessory is connected, it carries on, otherwise loop does not do anything.

The loop has two timer mechanisms that perform actions every half second and every minute. The half-second timer calls a function to update the current average reading and then send an update message to the phone. This message has an identifier "R" that is used when it is received at the phone, along with the smoothed value.

The one-minute timer sends a logging message ("L" identifier) and the count that has accumulated during the previous minute. Once the message has been sent, the count is reset to 0.

As we mentioned before, we use interrupts to call the "eventInterrupt" function every time the GM tube receives an event. This function updates the "instantaneousCPM" variable, increments the minute-wise count, and then sends an event ("E") message. The event message does not send an updated reading, but just notifies the phone so it can make the classic Geiger counter click and flash the radiation symbol.

The "sendMessage" function formats the message and then sends it. The message is packed into four bytes. The first byte is an identifier for all messages coming from this accessory (0×04). The second byte contains the flag, which will either be "R", "E", or "L", depending on whether the message is a "Reading," "Event," or "Log" message. The final two bytes are the 16-bit integer for the reading.

The Android App

The Android app is too big to be listed in full here, so if you are interested, you are encouraged to download the source from www.duinodroid.com. Let's now have a look at the key parts of the code.

One nice feature of the Android ADK is being able to cause the phone to launch an app in response to an Arduino accessory being plugged in. This is entirely accomplished by configuration in the AndroidManifest.xml file for the project.

This is the key section of the manifest.

```
<activity android:name="org.simonmonk
    .geiger.UsbAccessoryActivity"
          android:label="DroidGeiger"
          android:taskAffinity=""
          android:launchMode=
          "singleInstance">
  <intent-filter>
    <action android:name="android
    .hardware.usb.action.USB_
    ACCESSORY_ATTACHED"/>
  </intent-filter>
  <meta-data android:name="android
   .hardware.usb.action.USB_ACCESSORY_
   ATTACHED"
   android:resource="@xml/accessory_
   filter"/>
</activity>
```

The accessory filter determines which accessories qualify to launch this app. The filter for this app is as follows:

```
<resources>
    <usb-accessory manufacturer=
    "Simon Monk" model="DroidGeiger"
    version="1.0"/>
</resources>
```

In other words, only launch the app if the manufacturer, model, and version sent from the Arduino device all match those values set.

Let's now follow the path that a message coming form the Arduino would take.

Looking in the "run" method of the DroidGeigerActivity class, you will find the following fragment of code:

```
case 0x4:
  if (len >= 3) {
    Message m = Message.obtain
    (mHandler, MESSAGE_TEMPERATURE);
    char flag = (char)buffer[i + 1];
    int cps = composeInt(buffer[i + 2],
    buffer[i + 3]);
    m.obj = new GeigerMsg(flag,
    countsPerQuarter);
    mHandler.sendMessage(m);
}
i += 4;
break;
```

This method decodes the raw message coming from the USB connection and creates a message object to send to a Handler. Handlers are Arduino's way of allowing you to interact with the UI thread.

As shown in the following code snippet, the Handler will invoke the handleGeigerMessage method on the InputController class:

```
public void handleGeigerMessage
    (char flag, int reading) {
  if (flag == 'E') {
    mRadiationImage.setVisibility
    (ImageView.VISIBLE);
    mp.start();
  }
  else if (flag == 'R') {
    mRadiationImage.setVisibility
    (ImageView.INVISIBLE);
    mTemperature.setText("" + reading);
  }
  else if (flag == 'L') {
    String logText = mLogView
    .getText().toString();
    String timeFormatted = (String)
    DateFormat.format("hh:mm", new
    Date());
    mLogView.setText(logText + "\n" +
    timeFormatted + "\t\t\t" +
    reading);
  }
}
```

This makes the necessary changes to the user interface depending on the message flag.

Summary

There are many ways that this project could be extended. For instance, the Android app could, with a little work, be made to send the logged data over the mobile network, allowing remote monitoring.

It could also post readings to a web server.

This is a basic app and it is hoped that readers will take this further and do all sorts of interesting things with it.

In the next chapter, we look at another open accessory project: This time, a light show accessory—because the Evil Genius does love to party!

CHAPTER 3

Android Light Show

Bright flashing lights are great for disorienting your enemies. But even more importantly, the Evil Genius likes to dance, and can often be seen boogieing about the Evil Genius Lair to the latest disco beat, lights flashing away to provide atmosphere.

Occasionally, the minions are ordered to round up some guests for these parties, and the Evil Genius likes to impress such audiences with a demonstration of his Android phone–controlled light show.

Once the phone is docked into its light show Android Accessory (Figure 3-1), with a press of his finger the Evil Genius can adjust his fabulous light show.

The control application for the light show is shown in Figure 3-2.

The light show accessory uses three LED boards, each of which has 36 LEDs. The boards are each a different color: red, green, and blue. The brightness of each LED panel can be

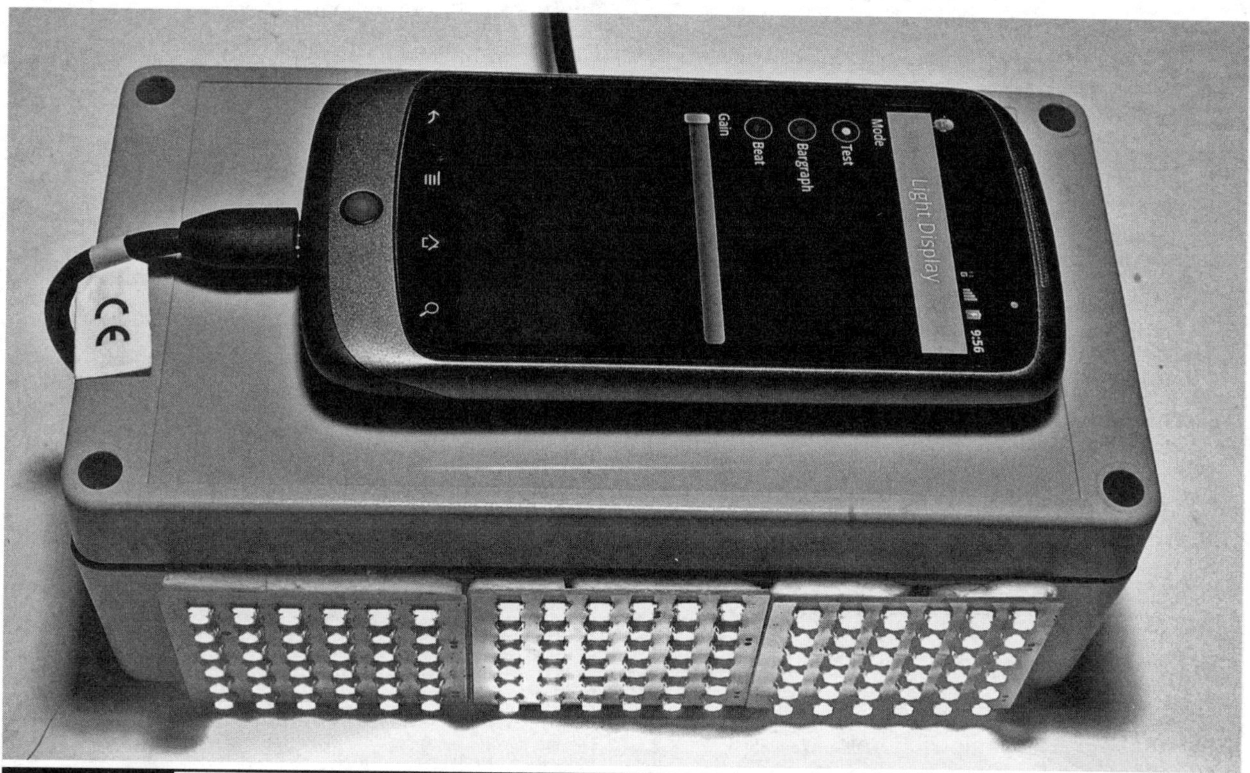

Figure 3-1 The light show accessory

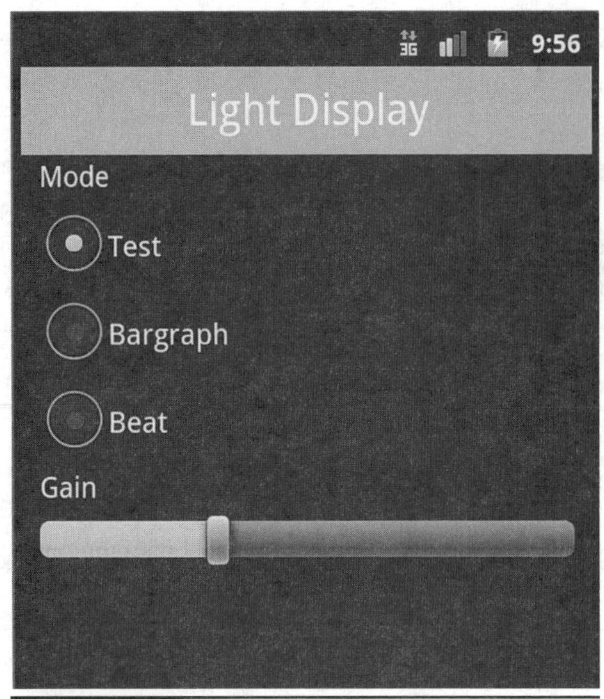

Figure 3-2 The light show app

controlled independently from the Android app that monitors the sound level through the phone's microphone and flashes the light accordingly.

Construction: The Droid Accessory Base

Like the Geiger counter, this is another Android Open Accessory project. However, this time, because we only have a few components aside from the Arduino, we are going to convert a USB host shield to be the basis of an off-board Arduino. In other words, we are going to use the Arduino to program the ATMega328 microcontroller and then transfer it onto an IC socket on the shield itself, meaning we can keep the Arduino for another project.

This is a bit like making our own Arduino on a USB host shield and just bringing the few Arduino pins we need for this project out onto a six-pin socket. It will give us three PWM pins and one general IO pin. The complete Droid Accessory Base is shown in Figure 3-3.

Figure 3-3 The Droid Accessory Base

We'll use this as the basis for this project and the next three that follow.

The schematic diagram for the project is shown in Figure 3-4.

As you can see from the schematic, we only need a minimum set of other components to allow the microcontroller chip to work on its own without the Arduino board. We need a crystal oscillator, a resistor, a couple of capacitors, and that's it.

What You Will Need (Droid Accessory Base)

The following Parts Bin is for the "Droid Accessory Base," as we will call this module. To make the light show, you will also need the other components listed in the section for constructing the project itself.

In addition to these components, you will also need the following tools in the Toolbox.

TOOLBOX

- Soldering equipment
- A computer to program the Arduino
- A USB-type A-to-B lead

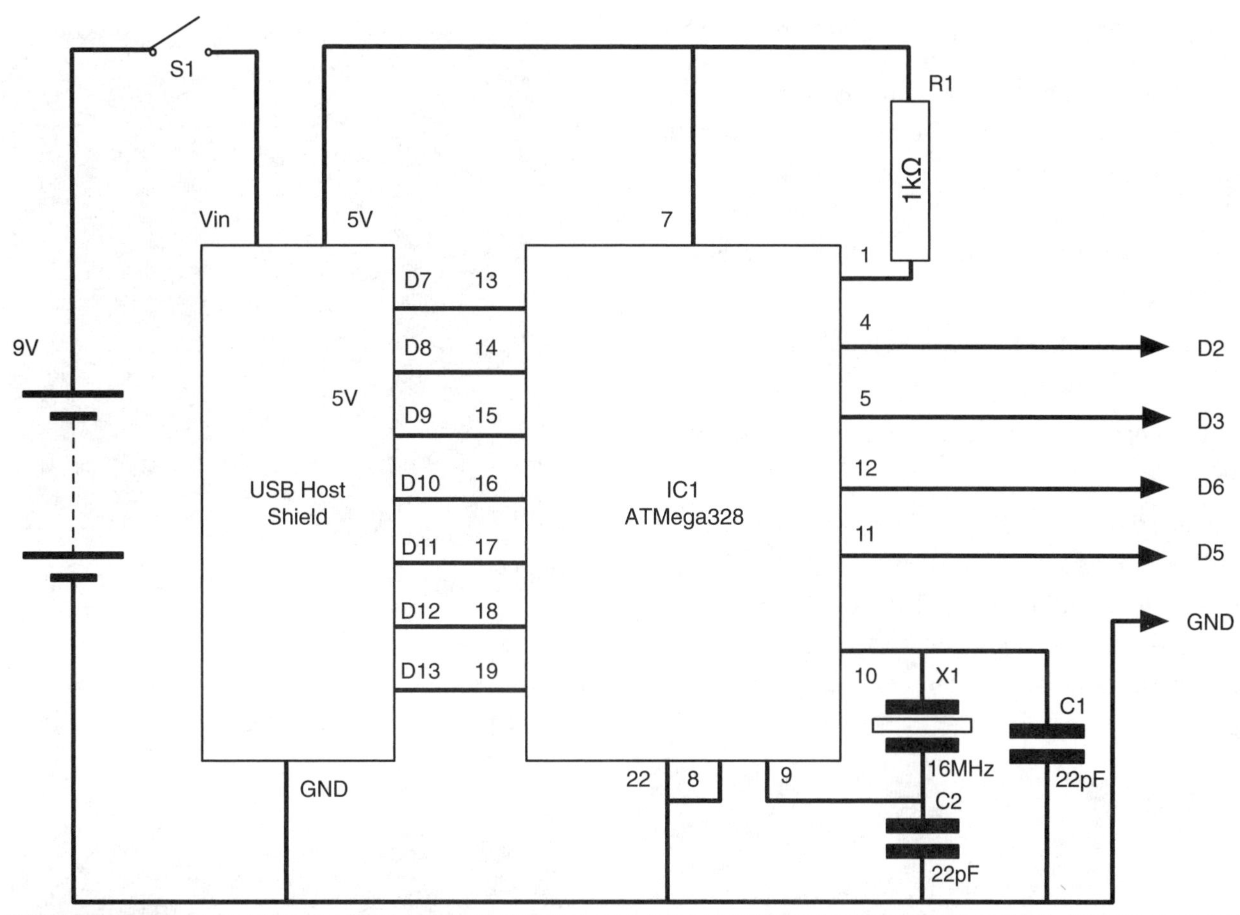

Figure 3-4 The schematic diagram

PARTS BIN			
Part	**Quantity**	**Description**	**Source**
Arduino Uno	1	Arduino Uno board (for programming)	www.arduino.cc
USB shield	1	Arduino USB Host Shield	Sparkfun: DEV-09947
Microcontroller	1	ATMega328 with bootloader	Sparkfun: DEV-10524
R1	1	1kΩ 0.5W metal film resistor	Farnell: 9339779
C1, C2	2	22pF ceramic capacitor	Farnell: 1600966
X1	1	16-MHz crystal	Farnell: 1611761
IC socket	1	28-pin DIP IC socket	Farnell: 1824463
Socket	1	6-way header socket	
Box	1	Small project box	RS/local electronics store
Switch	1	SPST miniature toggle switch	Farnell: 1661841

Step 1. Solder the IC Socket in Place

In this project, most of the wiring is going to be on the underside of the board, so rather than start with the wire links, we can solder the IC socket into place. This has the great advantage of making it easier to locate the positions for link wires relative to the IC socket.

The wiring plan for the prototype board is shown in Figure 3-5.

The board with the IC socket in place is shown in Figure 3-6.

Place the IC socket so that the notch indicting the location of pin 1 is as shown in Figure 3-5. You only really need to solder the IC socket pins that are used. It is also a good idea to solder one from each corner first to keep the socket in place.

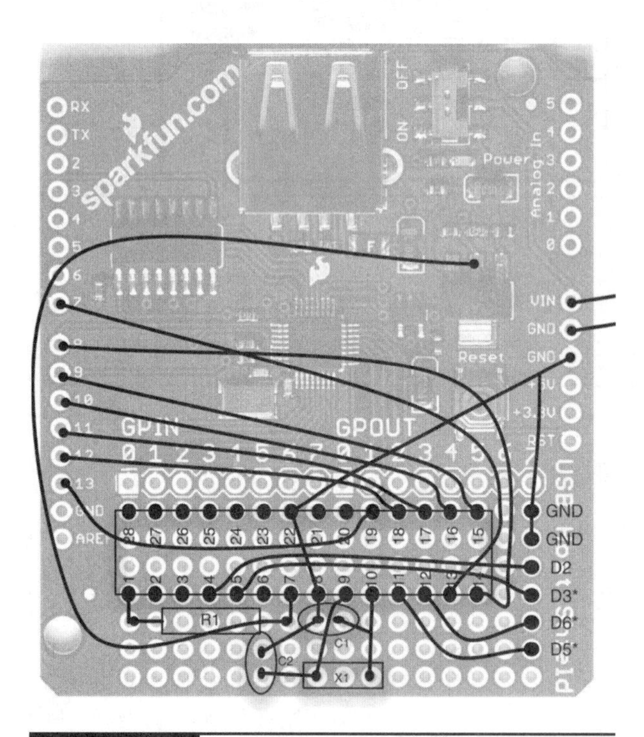

Figure 3-5 The wiring diagram

Step 2. Attach Link Wires for the USB Shield

Next, attach the link wires between the pins that would attach to the Arduino (if we were using one)

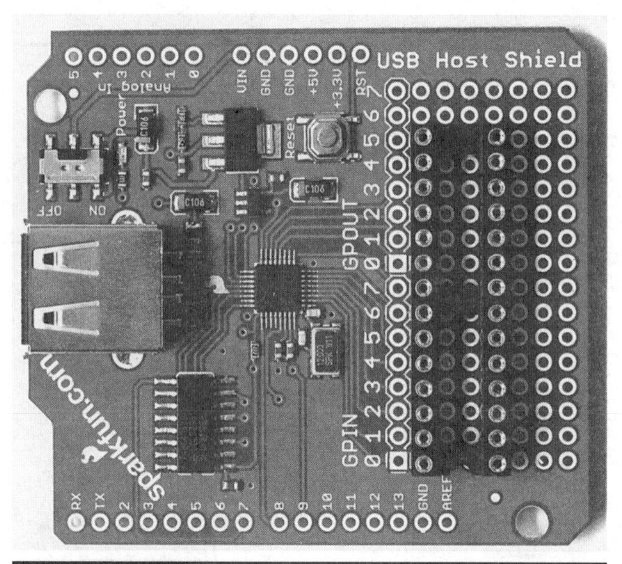

Figure 3-6 The shield with the IC socket in place

to the IC socket, where the Arduino's processor will eventually reside. Use Figure 3-5 and Figure 3-7 as a reference.

Figure 3-7 The shield with link wires attached

This is a little fiddly. Cut the wires to the right length and strip off a short length of insulation from each end, then solder each wire into place, making sure there are no solder bridges to other pins of the IC.

Step 3. Fit the Crystal and Other Components

We now return to the top side of the board and solder in the crystal X1, the two capacitors, and the resistor. See Figure 3-8.

Figure 3-8 The top side components in place

Solder the component leads at the hole through which they enter the board, but do not snip off the excess leads. This will allow you to bend them into position so they can form the connections between the components. Use Figure 3-5 as a reference when doing this.

We can also take the opportunity to solder the six-way socket, to which we will attach our accessories.

The end result should look something like Figure 3-9.

Step 4. Fit the Remaining Links

We can now return to the underside of the board and make the remaining links to the six-pin connector and the ground connection.

The end result is shown in Figure 3-10.

Figure 3-9 Connecting the components underneath the board

Figure 3-10 Fitting the remaining link wires

Step 5. Connect the 5V Supply

Since we do not have an Arduino board to supply us with the 5V, we must get the 5V supply from the shield. We get this by carefully soldering a lead between the middle pin of the large voltage regulator and the side of R1 that connects to pin 7 of the IC (Figure 3-11).

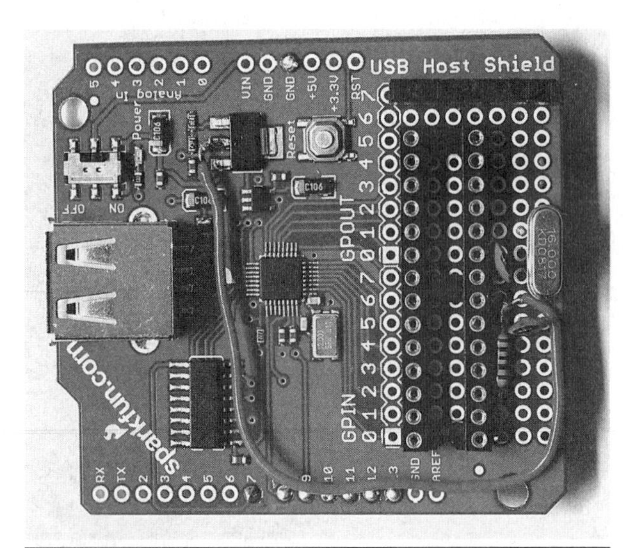

Figure 3-11 Attaching the 5V supply

Step 6. Connect the Switch and Power Leads

All that remains for us to do to complete the Droid Accessory Base is to attach the switch and power leads (Figure 3-12).

The Droid Accessory is used in both battery- and AC adapter–powered projects. So for now, we will attach a switch and two leads. Those leads can then either be connected to a battery clip, or as is the case with the light show project that follows, a power socket.

Step 7. Test

There are a lot of connections, all close together in this project, so take some time to inspect the board and make sure you have made all the connections as described in Figure 3-5 and that there are no unwanted solder bridges.

We will eventually use the light show project to test this, but we can get a basic sanity check by installing the Droid Geiger sketch onto an Uno and then carefully easing the microcontroller IC off the board and fitting it onto the IC socket on the shield. Be very careful not to bend the leads on the IC socket, because it is easy to snap them off.

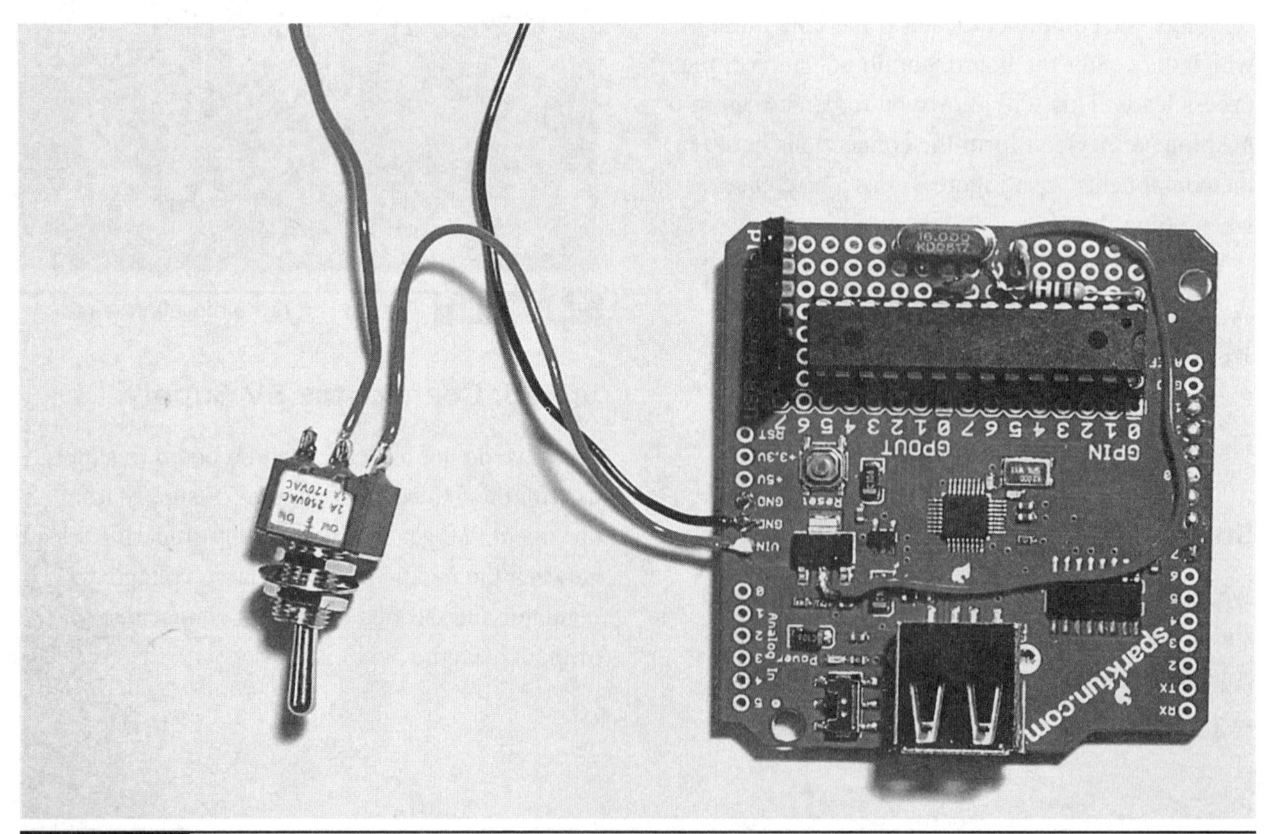

Figure 3-12 Attaching the switch and power socket

Also, make sure you put the IC the correct way around, with the notches on the IC and the IC socket lining up.

Hold or clip the connections from a 9V battery to the power leads from the Droid Accessory Base and plug the USB lead into your fully charged Android phone. If you installed the Droid Geiger app for the project in Chapter 2, it should be launched, although you will obviously not get any readings.

If the Droid Geiger app is not installed, you should get a popup message telling you that you can get it from this book's web site.

In both cases, this is a good sign.

Construction: The Light Show Project

Now that we have made our Droid Accessory Base, the rest of this project (and the next three projects) are relatively straightforward.

The schematic diagram for the project is shown in Figure 3-13.

The project uses the three PWM pins of the Droid Accessory Base. It uses pin D5 to control the red LED module, D6 to control the green, and D3 the blue. Each of these pins drives a MOSFET transistor to control the power to the LED modules since they use too much power to be driven directly from the microcontroller.

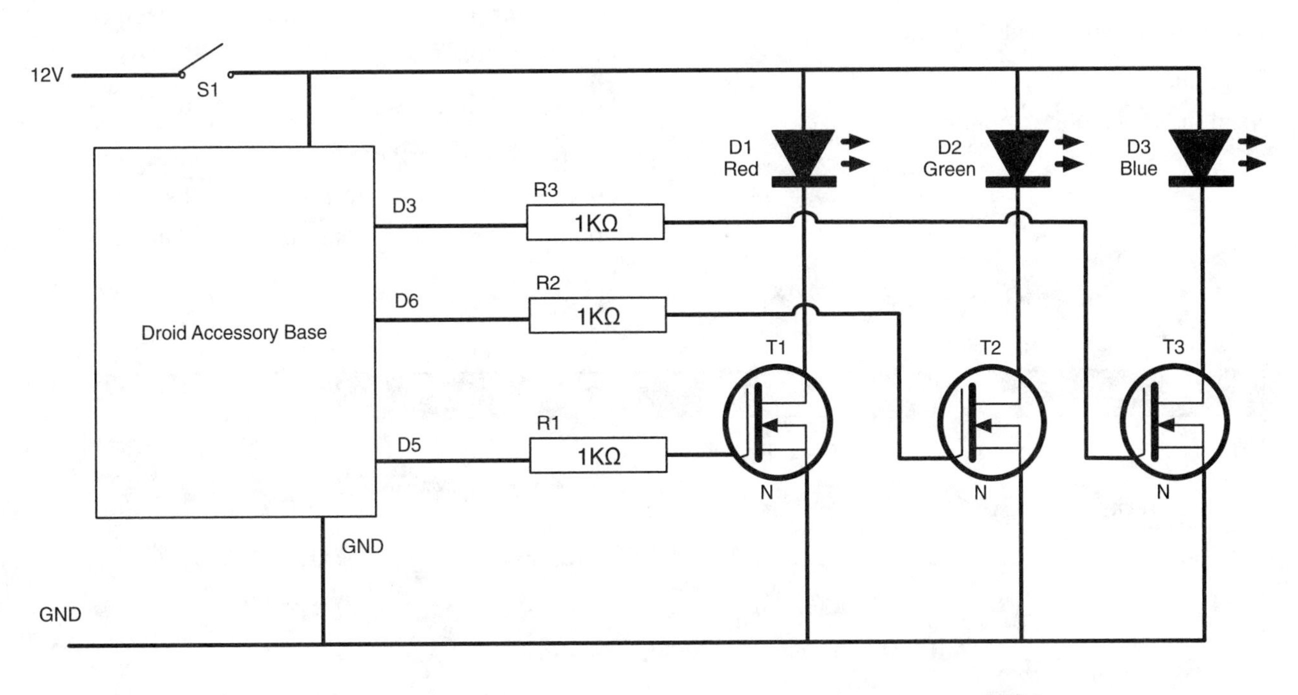

Figure 3-13 The schematic diagram for the light show itself

What You Will Need (Light Show)

The following Parts Bin is for the actual light show accessory. It assumes you have already built the Droid Accessory Base described earlier in this chapter.

The LED modules are each made up of 36 high-brightness surface-mount LEDs. They are designed to be powered from 12V. They have handy self-adhesive backs that we will use to stick them to the box. You can, however, use any 12V DC LED modules for the lighting, and the transistors should cope easily with up to 10 watts per channel, although be aware that you would need a much bigger power supply to cope with this. If you do not pick LED modules with 0.1-inch header pins as connectors, you will need to use a different connector arrangement.

In addition to these components, you will also need the tools in the Toolbox.

TOOLBOX

- Soldering equipment
- An electric drill and assorted drill bits
- Small self-tapping screws
- A computer to program the Arduino
- A USB-type A-to-B lead

Step 1. Make a Droid Accessory Base

See earlier in this chapter.

Step 2. Cut the Perfboard to Size

Cut a small piece off the perfboard that is 20 holes by 10. The best way to do this is to score the board with a craft knife and then break it over the edge of your bench.

Figure 3-14 shows the perfboard cut to size.

The idea of perfboard is that you push components through from the top and solder them up underneath, connecting them together with their leads and, where necessary, extra lengths of wire.

PARTS BIN

Part	Quantity	Description	Source
Droid Accessory Base	1	*See the start of this chapter*	
R1-3	3	1kΩ 0.5W metal film resistor	Farnell: 9339779
T1-3	3	FQP33N10 MOSFET transistors	Farnell: 9845534
D1	1	Red LED Module *(see later)*	eBay
D2	1	Green LED Module *(see later)*	eBay
D3	1	Blue LED Module *(see later)*	eBay
Perfboard	1	20 by 10 holes	Farnell: 1172145
Socket header	1	Six-way 0.1-inch header socket strip	Farnell: 1218869
Screw terminal	1	Two-way 0.2-inch (5mm) spacing	Farnell: 1641932
Project box	1	Large project box	Local electronics store
Power socket	1	2.1mm power socket	Farnell: 121703
Power adapter	1	12V 1.5A power supply	Local electronics store

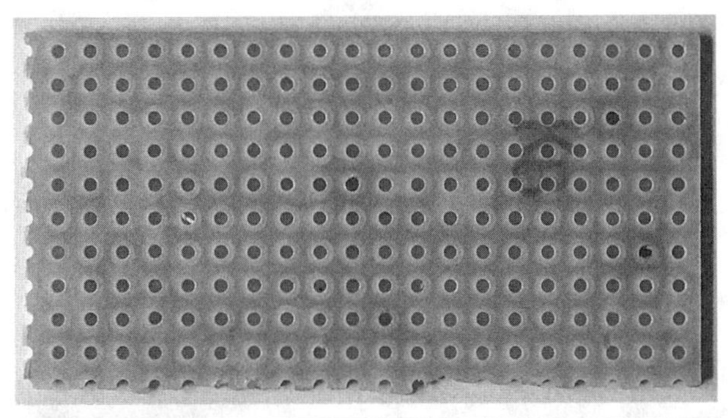

Figure 3-14 The perfboard cut to size

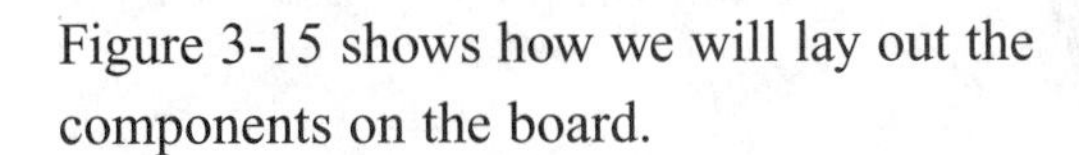

Figure 3-15 shows how we will lay out the components on the board.

Step 3. Attach the Socket, Screw Terminal, and Transistor

The tricky bit in using perfboard is getting everything to hold together. A good starting point is to place the header socket, the screw terminal, and one of the transistors onto the board (Figure 3-16).

Figure 3-17 shows the underside of the board. You can see that the leads of the transistor just bend far enough to reach one contact of the screw terminal and one of the header sockets.

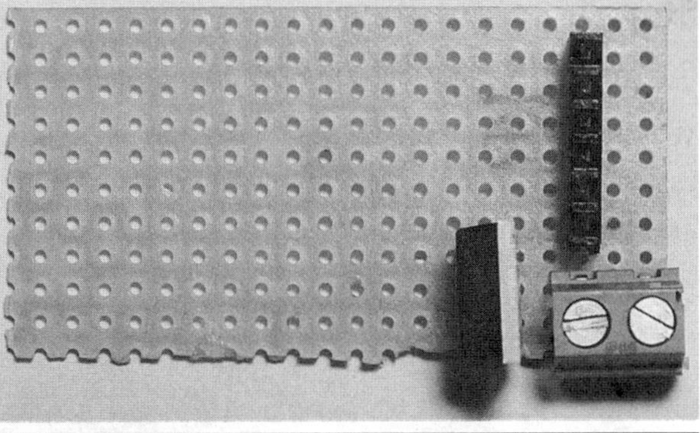

Figure 3-16 Attaching the first components

Figure 3-17 The underside of the board with the first components attached

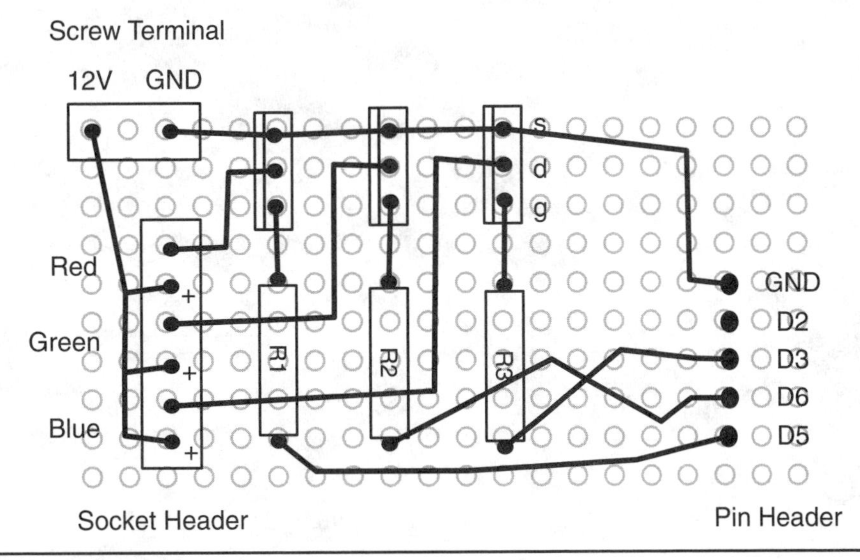

Figure 3-15 The perfboard layout

Step 4. Solder the Remaining Components

We can now attach the rest of the components. It is easiest just to put all the remaining transistors and the resistors into their correct places on the board and then bend the component leads in the right directions for them to be connected together, and then solder them up.

Figures 3-18 and 3-19 show the top and bottom of the board after everything is soldered together.

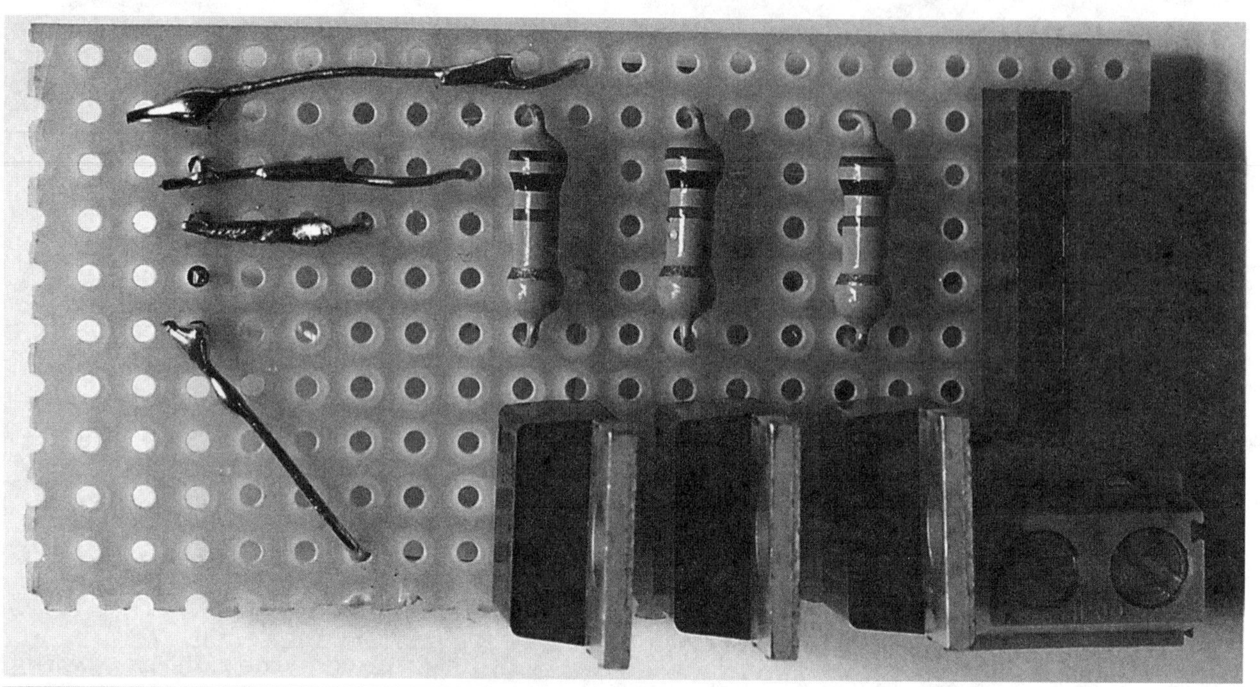

Figure 3-18 The top side of the completed board

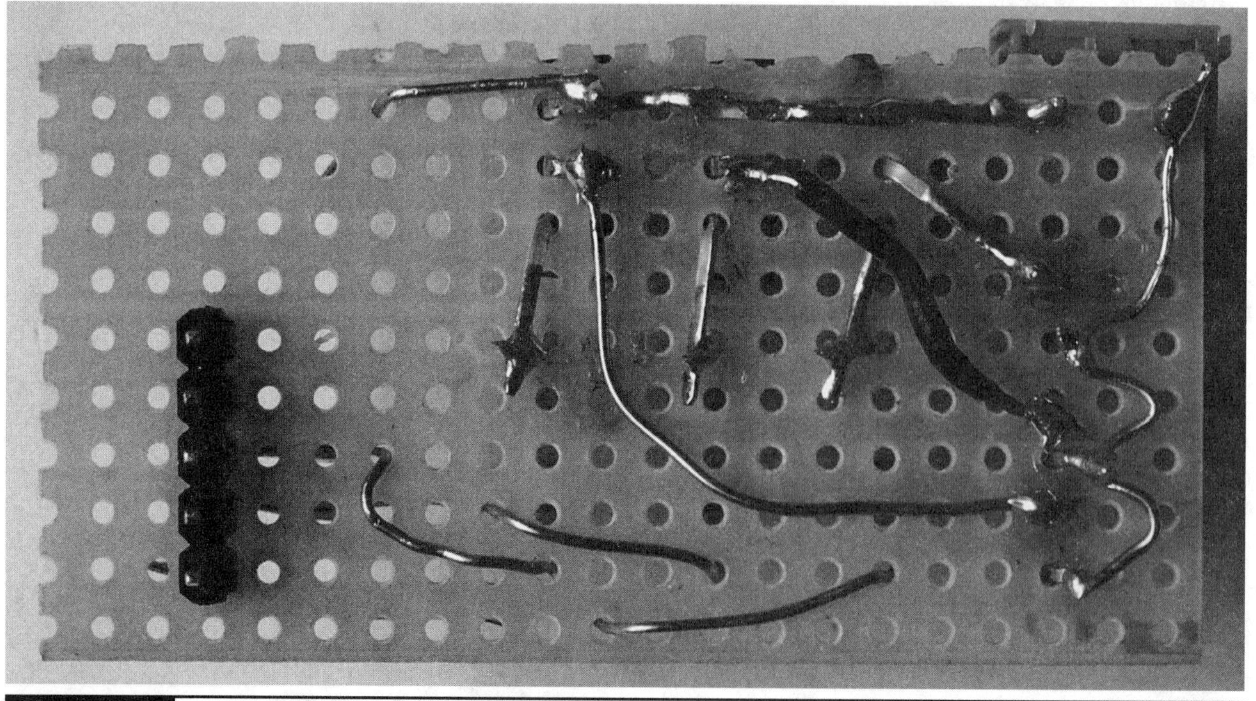

Figure 3-19 The bottom of the completed board

The pin header only uses four of the connections, one for ground, and three connections that go to R1 to R3. You can see from Figure 3-18 how the resistor leads are threaded through the perfboard, first down and then back up, to make a connection to the pin header on the top of the board. Only the nearest resistor to the header is likely to have a lead long enough to reach the pin header. The other two will need to be extended with some wire.

Notice the zigzagging wire shown on the right-hand side of Figure 3-19, which connects the positive supply terminal to the three positive connections for the LEDs. It is best to bend the wire to the correct shape and then solder it into place.

Step 5. Connect Everything Together

Figure 3-20 shows how everything is connected together.

Be very careful to get the connectors to the LEDs the correct way around.

As you can see from Figure 3-20, another wire has been soldered from the switch and screwed into the positive terminal on the perfboard.

We also need to take the flying leads from the Droid Accessory Shield and attach a 2.1mm power socket so we can connect our 12V DC adapter to the board (see Figure 3-20).

Figure 3-20 Connecting everything together

Step 6. Install the Arduino Sketch

To get the sketch onto the microcontroller, we'll first program the chip and then carefully take it off the Arduino board and fit it onto the socket on the Droid Duino Base.

The sketch uses the same libraries as the projects in Chapters 2 to 5, so if you have not completed one of these chapters, refer to the instructions in Chapter 2, Construction Step 6 ("Install the Open Accessory Libraries").

The sketch is called "ch03_light_show" and is included in the zip file containing all the sketches in the book that can be downloaded from www.duinodroid.com. See Chapter 1 for details.

Once the sketch is installed on the microcontroller, carefully remove the microcontroller chip, levering each end up a little at a time so as not to bend the pins. Then seat it onto the Droid Accessory Base, making sure you have it the right way around.

Step 7. Install the Android App

If your phone has an Internet connection, you can skip installation of the Android app and wait until the phone's Open Accessory software asks for it the first time you connect the light show accessory. Otherwise, visit www.duinodroid.com and download the APK installer.

If you have any trouble with this, read through Chapter 1, Construction Step 10.

Step 8. Test

The time has come to test our project. The sketch starts the accessory in its "test" mode, so even without connecting our phone, we can see the LEDs in action, by plugging in the 12V supply.

Next, we can plug in the phone. This project also acts as a charging station for the phone, so while the lights are flashing, the phone will also be charging.

When the phone is plugged in, it should launch the Duino Light Show app.

Step 9. Box the Project

The project fits into a project box big enough to act as a base for the three self-adhesive LED panels. The phone is attached with a USB lead fitted inside the case.

Figure 3-21 shows the box drilled with holes for the switch, the USB lead, the power socket, and the three holes for the LED modules.

Now, attach the shield to one of the pillars in the plastic box with a small self-tapping screw, and fit the switch and power socket (Figure 3-22).

Finally, we can fix the self-adhesive LED panels to the side of the box and thread the wires through to connect to the pin headers.

Make sure the LEDs are connected the right way round, and that the leads are long enough to reach the connector before you remove the self-adhesive pads (Figure 3-23).

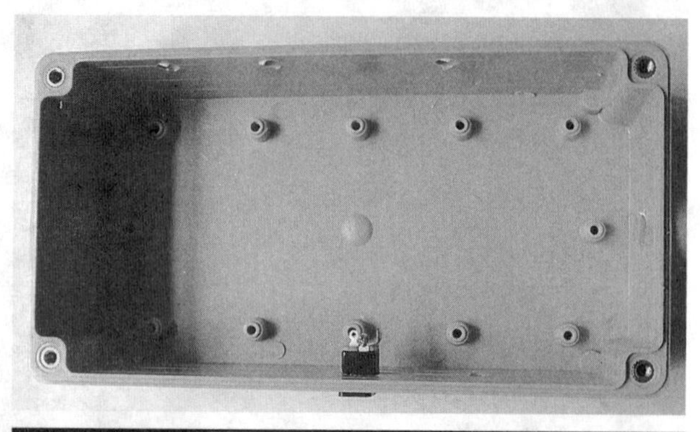

Figure 3-21 The project box, drilled

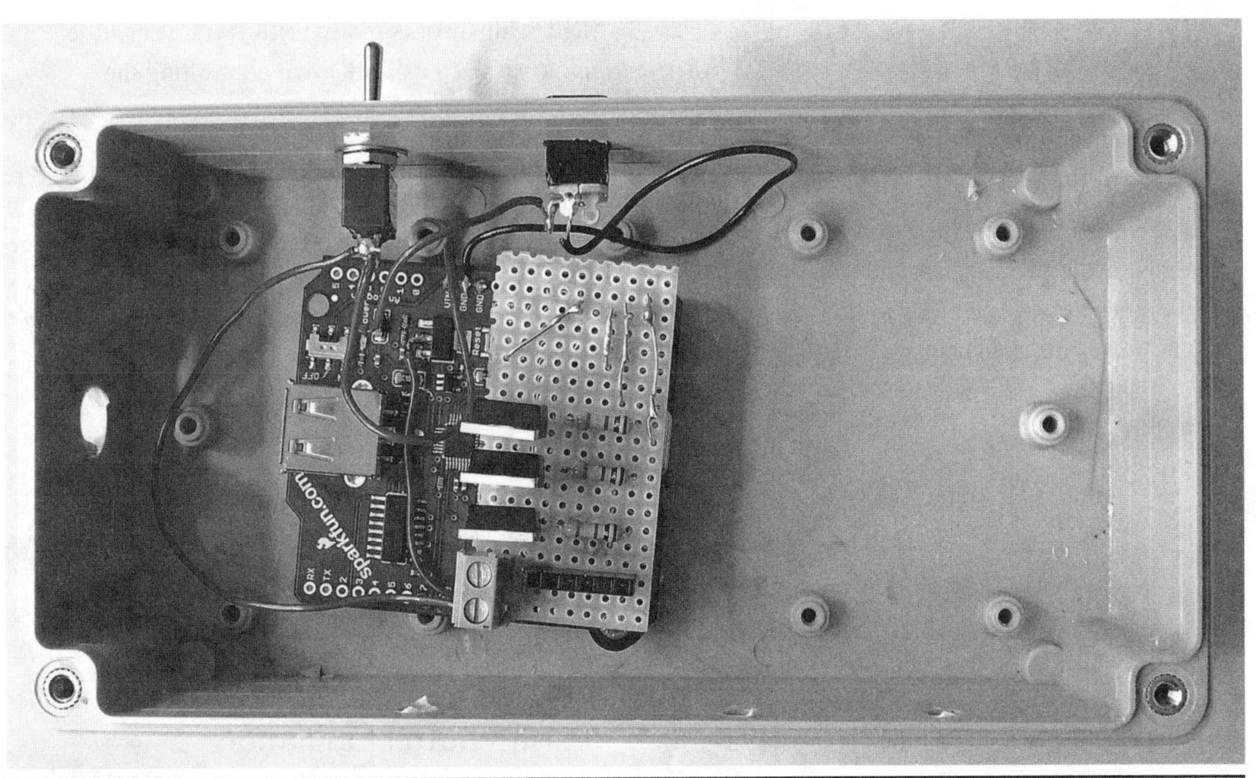

Figure 3-22 Attaching the switch and power socket

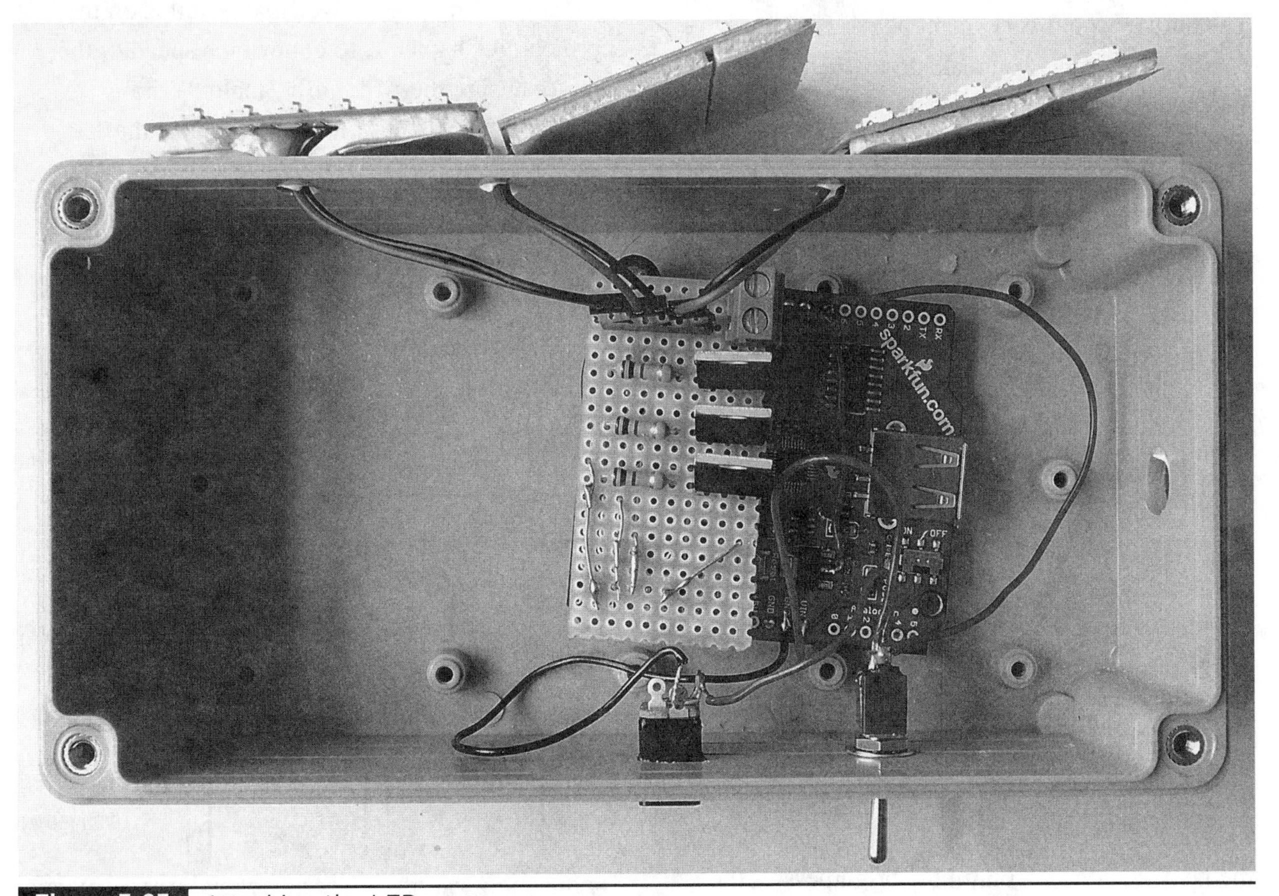

Figure 3-23 Attaching the LEDs

Using the Project

The project has three modes. It has the test mode that just cycles the LEDs even if no phone is connected to the accessory.

Once you put it into "Bargraph" or "Beat" mode, the device becomes sound-sensitive. It uses the phone's microphone and will therefore be responsive to both sounds played by the phone's Music player and external sounds. So you might like to plug the audio jack of the unit into an amplifier.

The sensitivity of the unit is adjusted using the "gain" slider.

Theory

The software for this and all the other projects in this book are provided as open source and you are encouraged to make your own improvements. The author would very much like to hear about any improvements you make to the software. Ways to contact the author can be found at www.duinodroid.com, where you will also find all the source code for both Android and Arduino.

In this section, we'll have a quick look at the Arduino sketch. But before that, we'll explore how to use MOSFETs to turn the LEDs on and off, and quickly discuss pulse width modulation.

MOSFETs

MOSFETs (Metal Oxide Semiconductor Field Effect Transistors) are a type of transistor that is excellent for switching high currents. When they are turned fully off, they have a huge resistance, and when they are on, they have a very low resistance. This means that they generate very little heat, when used as a switch that is either fully on or fully off.

Since we are using pulse width modulation (see the next section) to control the brightness of the LEDs, this works really well—with very little power wasted as heat when controlling the brightness of an LED.

They differ from more normal "bipolar" transistors in that they are controlled by voltage rather than current. In other words, it is the voltage at the "gate" connection of a MOSFET that determines whether it turns on or not. Only a tiny current flows into the gate, making them ideal for switching from low-current sources like the Arduino.

The MOSFETs we will use are called logic-level MOSFETs, because the gate voltage needed to turn the MOSFET on is less than 5V, and therefore the MOSFET can be turned on and off directly from the Arduino pins.

Pulse Width Modulation

Pulse Width Modulation (or PWM) is a technique for controlling power. We first used this on the project in Chapter 1, to control the speed of the motors using the Arduino's "analogWrite" function. We also use it here to control the brightness of the LED modules.

Figure 3-24 shows the signal from a PWM pin on the Arduino.

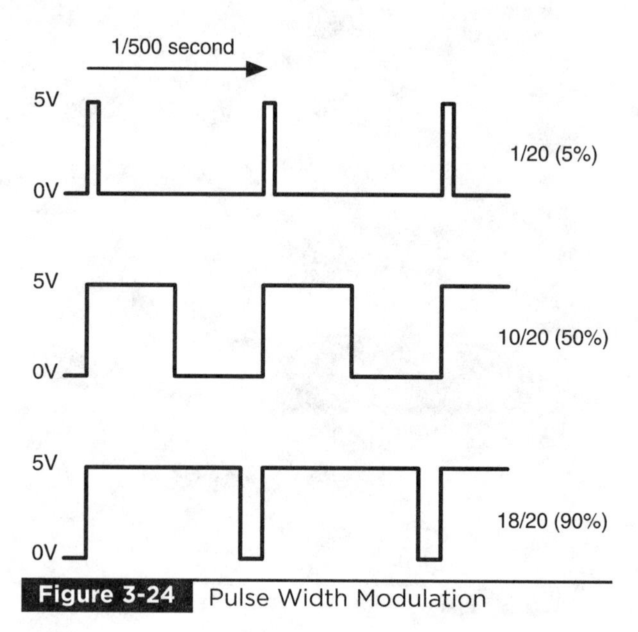

Figure 3-24 Pulse Width Modulation

The PWM pin is oscillating at about 500 Hz, with the relative amount of time that the pin is high varying with the value set in the "analogWrite" function. So, looking at Figure 3-24, if the output is only high for 5 percent of the time, then whatever we are driving will only receive 5 percent of full power. If, however, the output is at 5V for 90 percent of the time, then the load will get 90 percent of the power delivered to it.

When driving motors with PWM, the physical inertia of the spinning motor means it does not start and stop 500 times per second, but is just given a kick of varying strength every 500th of a second. The net effect of this is smooth control of the motor speed.

LEDs can respond much more quickly than a motor, but the visible effect is the same. We cannot see the LEDs turning on and off at that speed, so to us it just looks like the brightness is changing.

The Arduino Sketch

The sketch for this project is pretty straightforward. It differs from the previous project (the Geiger counter) because it is more concerned with receiving commands from the accessory. Whereas in the Geiger counter project, the communication was in the other direction.

```
#include <Max3421e.h>
#include <Usb.h>
#include <AndroidAccessory.h>

#define redPin 5
#define greenPin 6
#define bluePin 3

#define cycleTime 10

int red = 0;
int green = 85;
int blue = 170;

boolean randomMode = true;


AndroidAccessory acc("Simon Monk",
      "DroidLightShow",
      "Light Show Accessory",
      "1.0",
      "http://www.duinodroid.com/android",
      "0000000012345678");

void setup()
{
  Serial.begin(9600);
  pinMode(redPin, OUTPUT);
  pinMode(greenPin, OUTPUT);
  pinMode(bluePin, OUTPUT);
  acc.powerOn();
```

(continued)

```
}

void loop()
{
  byte msg[3];
  if (acc.isConnected())
  {
    int len = acc.read(msg, sizeof(msg), 1);
    if (len > 2 && msg[0] == 1) // set red
    {
      red = msg[2];
    }
    if (len > 2 && msg[0] == 2) // set green
    {
      green = msg[2];
    }
    if (len > 2 && msg[0] == 3) // set blue
    {
      blue = msg[2];
    }
    if (len > 2 && msg[0] == 4) // test mode on
    {
      randomMode = true;
    }
    if (len > 2 && msg[0] == 5) // test mode off
    {
      randomMode = false;
    }
  }
  if (randomMode)
  {
     changeColors();
  }
  showColors();
  delay(cycleTime);
}

void changeColors()
{
   red ++;
   if (red > 255) red = 0;
   green ++;
   if (green > 255) green = 0;
   blue ++;
   if (blue > 255) blue = 0;
}

void showColors()
{
  analogWrite(redPin, red);
  analogWrite(greenPin, green);
  analogWrite(bluePin, blue);
}
```

We start by including the libraries that we need. The Max3124e library is included as part of the USB host library you have already installed.

We then have a constant "cycleTime" that is used when the accessory is in test mode. This sets the time period for the delay between changes to the colors.

The next three variables contain the brightnesses for the red, green, and blue LEDs as a value between 0 (off) and 255 (full on).

After defining the variables, the line beginning with "AndroidAccessory" sets up the connection between the phone and the Arduino in the same way as the Geiger counter sketch.

The "setup" function starts up serial communications and initializes the output pins.

The loop function calls "acc.isConnected" and then checks for incoming messages using the "acc.read" function. The first byte of the message is the code for the type of message. A value of "1" sets the red brightness, "2" sets the green, and "3" the blue. The codes "4" and "5" turn the test mode on and off, respectively.

Most of the rest of the sketch is concerned with the "test" mode. This cycles the brightness of each of the LEDs from 0 to 255. Each LED has a different starting brightness.

The Android App

The Android app is too big to be listed in full here, so if you are interested, you are encouraged to download the source from www.duinodroid.com. We will have a look at the key parts of the code next.

Much of the Android app is similar to that of the Geiger counter app, so the curious might like to refresh their memories using the description of this in the "Theory" section of the previous chapter.

As with the Arduino sketch, the main difference is in the direction of communication. So for this project, it is the Android device that sends the message to the Arduino. The relevant section of code is shown next:

```
public void sendCommand(byte command,
  byte target, int value) {
    byte[] buffer = new byte[3];
    if (value > 255) value = 255;
    buffer[0] = command;
    buffer[1] = target;
    buffer[2] = (byte) value;
    if (mOutputStream != null &&
      buffer[1] != -1) {
        try {
          mOutputStream.write(buffer);
        } catch (IOException e) {
          Log.e(TAG, "write failed", e);
        }
    }
}
```

This code can be found in the class DroidSoundDisplayActivity. The message is constructed as a three-byte array that is sent to the USB output stream.

To sample the sound, we employ a much-passed-around GPL library used in the Blinkendroid project. The two classes used can be found in the package "org.cbase.blinkendroid.audio" and provide a neat callback mechanism that returns a section of sampled data. This is then analyzed by the class "Visualizer" that takes the sample and determines which messages to send to the Arduino.

Summary

That completes the second of our Open Accessory projects and the first using the Droid Accessory Base module we created. This will form the basis of the projects in the next three chapters.

CHAPTER 4

TV Remote

THE LIFE OF A MINION is a short brutal affair in which the minions must take pleasure where they can. One popular amusement of the Evil Genius' minions is to hide the remote control for the TV. Even severe punishments are not sufficient to deter the minions from indulging in this dangerous pastime.

In light of this, the Evil Genius decided to invent a project that would allow him to use his Android phone and a clever little accessory project to let him regain control of the TV whenever he wishes (see Figure 4-1).

Figure 4-1 The TV remote Android accessory

The Android control software for the project is shown in Figure 4-2.

The software allows eight buttons to be programmed that can send commands over IR.

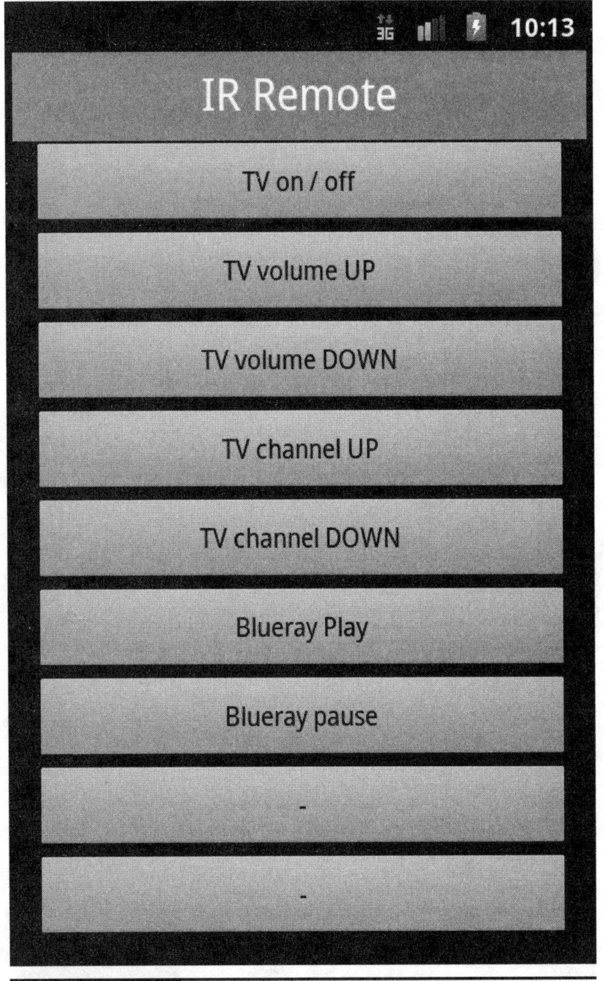

Figure 4-2 The TV remote Android app

These commands can be used for different appliances as well. Thus, the remote can be "trained" from a number of different remote controls so it will work as a "universal remote."

Construction

If you have constructed the Droid Accessory Base from Chapter 3, there is really very little construction to do since the few components can be attached to a small piece of stripboard.

The schematic diagram for the project is shown in Figure 4-3.

The project uses all of the IO pins on our Droid Accessory Base. It employs D3 to drive the IR transmitter, and the IR receiver chip uses pins D6 and D2 to provide the GND and positive supply to the IC. Pin D5 receives the IR signal from the receiver IC.

What You Will Need

In addition to an accessory-capable Android phone (Android 2.3.4 or later) and all the components required to make the Droid Accessory Base, you will need the components shown in the Parts Bin.

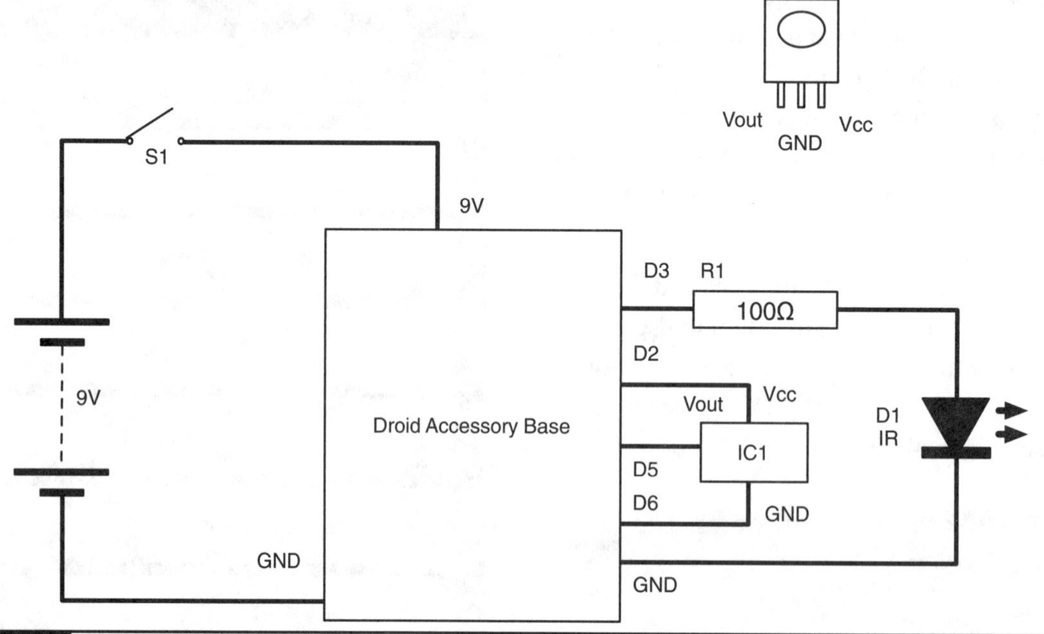

Figure 4-3 The schematic diagram

PARTS BIN

Part	Quantity	Description	Source
Droid Accessory Base	1	*See Chapter 3 for parts and building instructions*	
R1	1	100Ω 0.5W metal film resistor	Farnell: 9339760
R	Optional	1kΩ 0.5W metal film resistor	Farnell: 9339779
D1	1	5mm IR transmitter LED, 940nm	Farnell: 1020634
IC1	1	940nm IR receiver IC	Farnell: 4142822
Battery clip	1	9V battery clip	Farnell: 1183124
Stripboard	1	Eight strips each of six holes	Farnell: 1201473
Header pins	1	Six-way header pin	Farnell: 1097954
Case	1	Plastic case	Local hardware store

If you plan to make the noncharging USB lead, you will also need a 1kΩ resistor, and an old USB extension lead. See Chapter 2 for details on how to make this lead, which will prevent your mobile phone from draining your accessories' batteries.

As a longer-lasting battery alternative, you can use a holder that will accept six AA cells to supply the project at 9V.

In addition to these components, you will also need the following tools.

TOOLBOX

- Soldering equipment
- An electric drill and assorted drill bits
- Assorted self-tapping screws
- A computer to program the Arduino
- A USB-type A-to-B lead

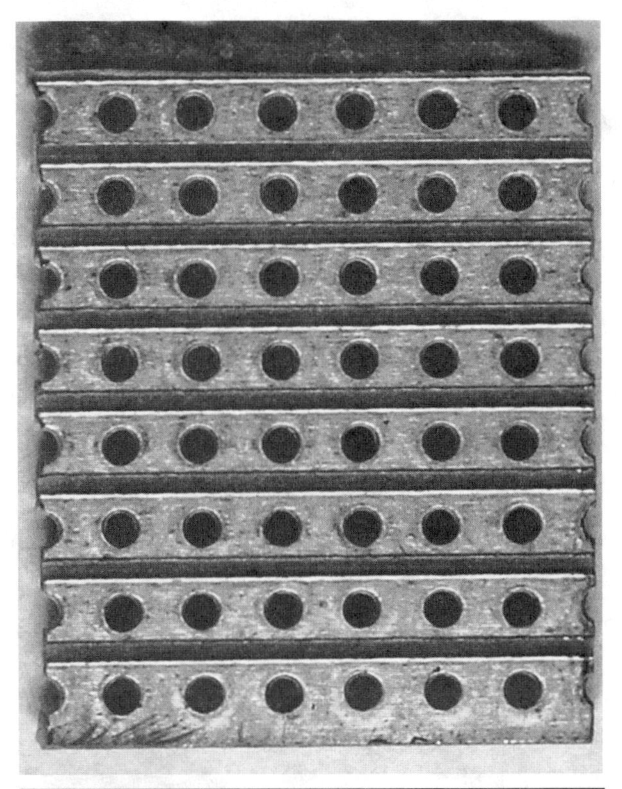

Figure 4-4 The stripboard cut to size

Step 1. Make a Droid Accessory Base

This is the unit described in Chapter 3 that is comprised of a USB host shield, an ATMega328 microprocessor programmed in your Arduino, and a few other components.

Step 2. Cut the Stripboard to Size

Cut the board to the right size. The neatest way to do this is to score the board heavily with a craft knife and then break it over the edge of a table. You can also use a strong pair of scissors, but the result will not be as neat (Figure 4-4).

The stripboard does not require any breaks to be made or links to be made. The layout is shown in Figure 4-5. Note that the layout is shown both from the copper and component sides of the board. It is shown this way because the header strip will emerge from the non-copper side of the board to fit into the Droid Accessory Base so that the board is fitted upside down onto the Droid Accessory Base, with the IR LED and IR receiver facing outwards.

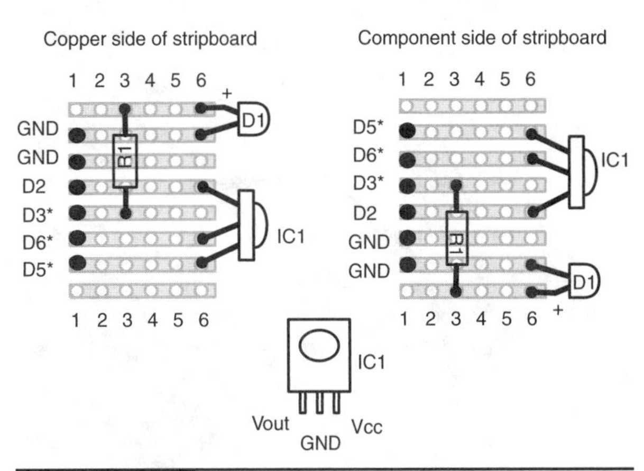

Figure 4-5 The stripboard layout

Step 3. Solder the Components

Only the three components and the header strip need to be soldered to the board.

Make sure the IR LED is the correct way around. The longer positive lead should be the one

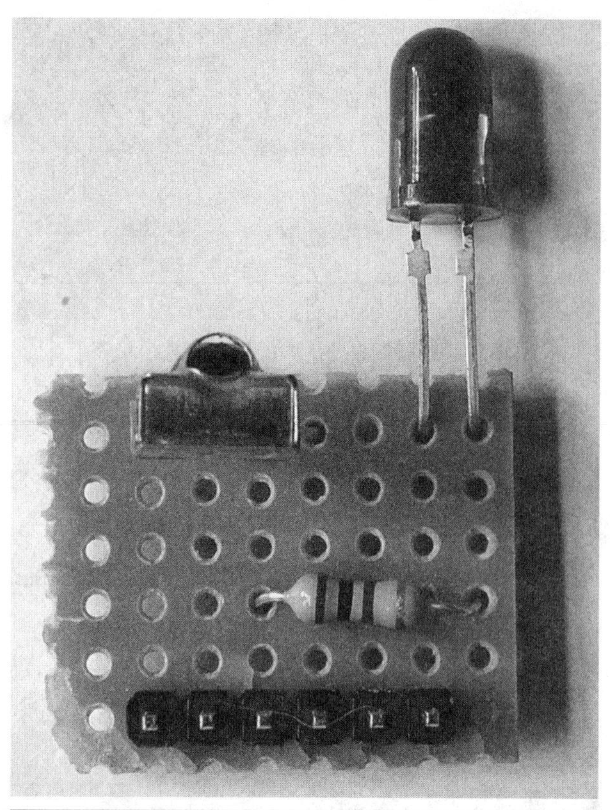

Figure 4-6 The completed stripboard

that is connected to one of the resistor's leads (Figure 4-6).

Step 4. Wire Up

The stripboard is shown attached to the base in Figure 4-7.

You will also need to solder a battery clip to the Droid Accessory Base and switch as shown in Figure 4-7.

Step 5. Install the Arduino Sketch

To get the sketch onto the microcontroller, we are first going to program the chip and then carefully take it off the Arduino board and fit it onto the socket on the Droid Duino Base.

This sketch only uses the Android libraries. See Step 6 of Chapter 2 for instructions on installing the Android libraries into your Arduino software.

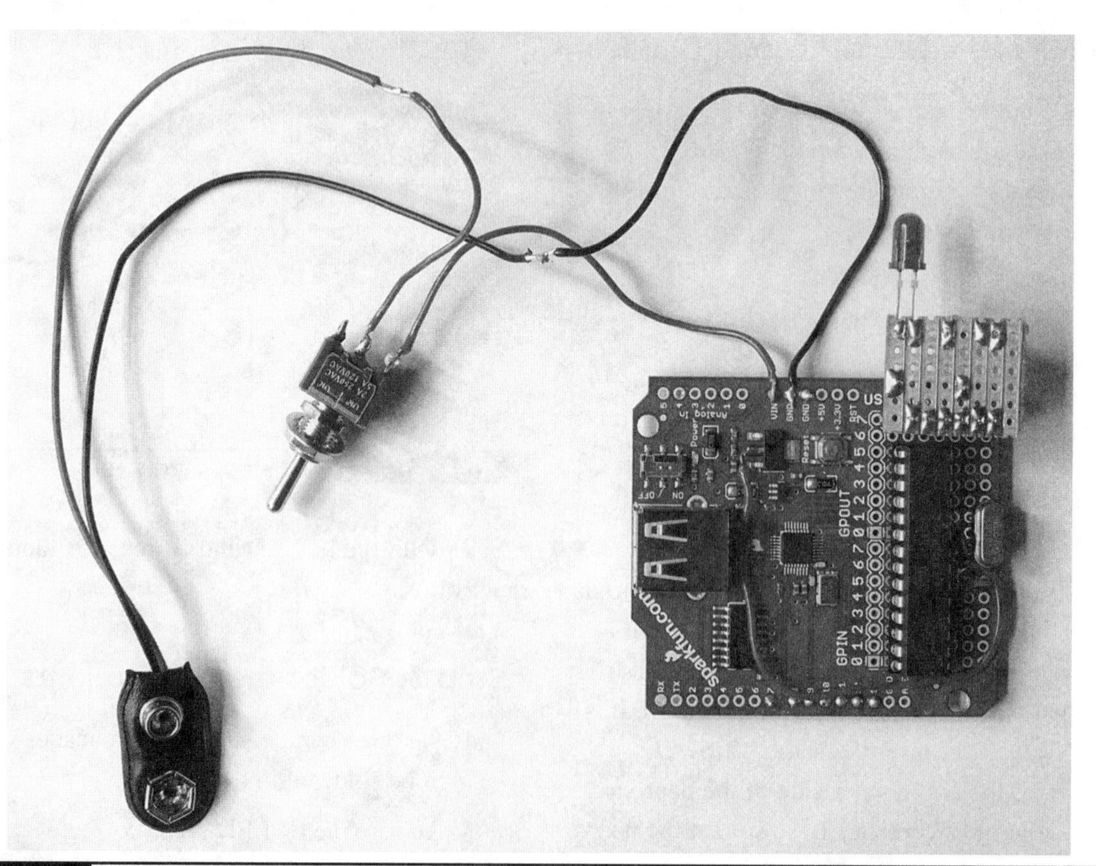

Figure 4-7 The stripboard attached to the base

We can now install the sketch (ch04_tv_remote) onto the Arduino board. Then, with the power disconnected, carefully remove the IC from the board and fit it into the socket on the Droid Duino Base, making sure it's the right way around.

Step 6. Install the Android App

If your phone has an Internet connection, you can skip installing the Android app and wait until the phone's Open Accessory software asks for it, which is when you first connect the TV remote accessory. Otherwise, visit www.duinodroid.com and download the APK installer for this project.

If you have any trouble with this, go back to Chapter 2, Step 10.

Step 7. Test

It is always a good idea to test your project before you fit everything into a box. So, connect everything up, as shown in Figure 4-7.

Turn on the switch and it should cause the phone to either launch the Droid TV Remote app, or if it is not installed, to display the URL from where it can be downloaded.

Note also that for this project the Duino Droid Base has a 9V battery clip connected to the power leads.

In this project, the Android phone does little more than act as a keyboard, passing through commands to the Arduino. Even the training of the remote and the storing of the codes is done on the Arduino device. So, we need to start by training the remote with a command from one of our conventional remotes. To do this, we need to first select a button that we want to program (by pressing it), then click the Menu button and select the Program option. Hold the remote near the IR receiver and press a button. If the code is received, then the app should notify you the code has been saved.

Whenever you program the remote in this way, it will program the last button that was pressed.

Now aim the remote at the device that the original remote controls and press the first button on the remote. This should have the same effect as the original remote.

Turning the device to standby is always a good button to start with. Note that the IR transmitter is not as bright as a regular remote, so you will have to get closer to the device than with the original remote.

You can often see the IR LED if you view it through a camcorder or digital camera, so if you are not sure whether the remote is sending a code, you can check it using this method.

Step 8. Box the Project

The process of boxing this project is very similar to the two previous projects. First, lay out the components in the box (Figure 4-8), and then work out what holes you will need where and how big they should be.

You will need holes for:

- The USB cable
- The screws in the base of the box to hold the board in place
- The switch
- The IR receiver and transmitter

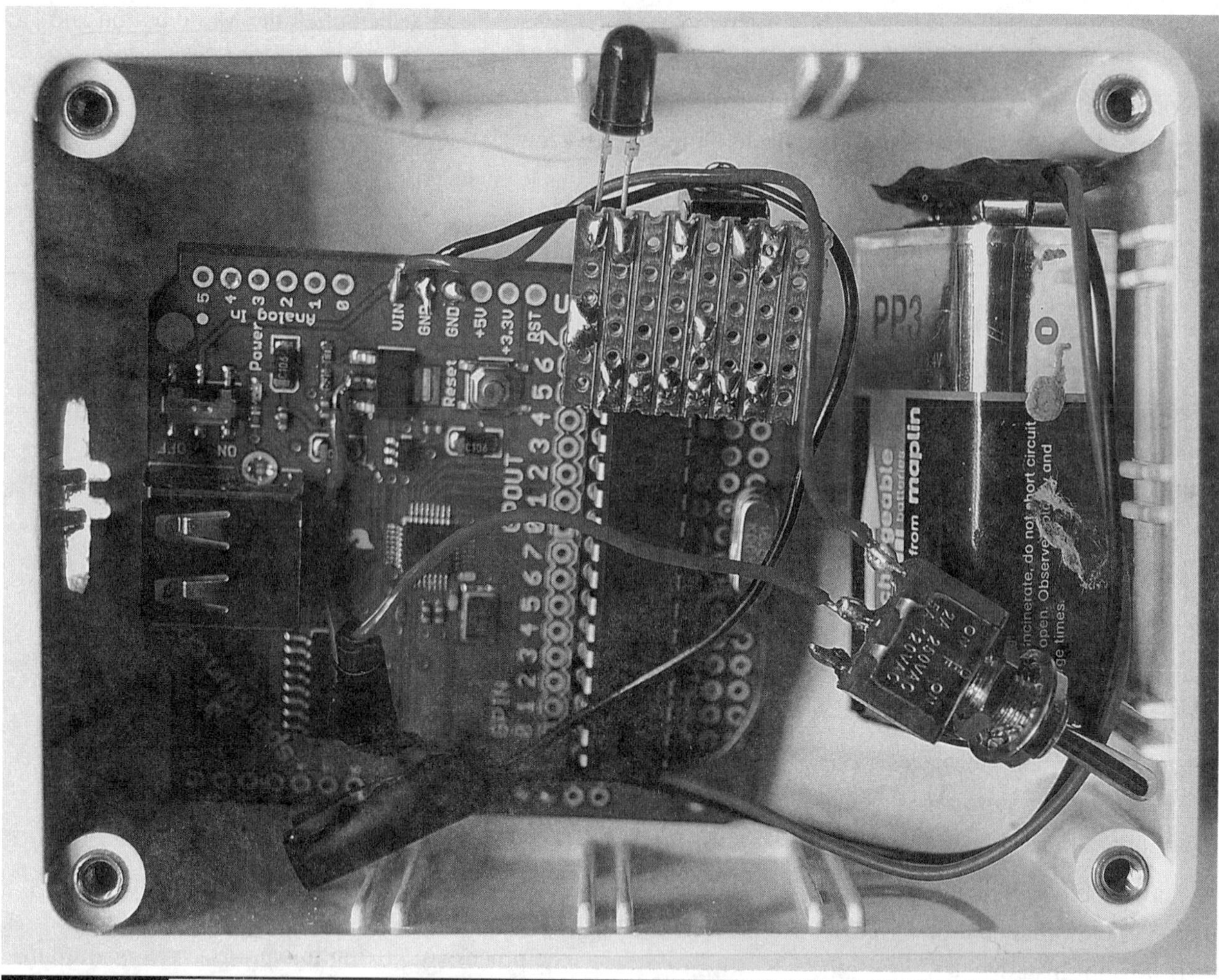

Figure 4-8 Laying out the components in the box

Using the Project

You can change the label of any of the buttons by doing a "long touch" of the button (pressing and holding it for a second or two). This opens the settings screen where the name can be edited (Figure 4-9).

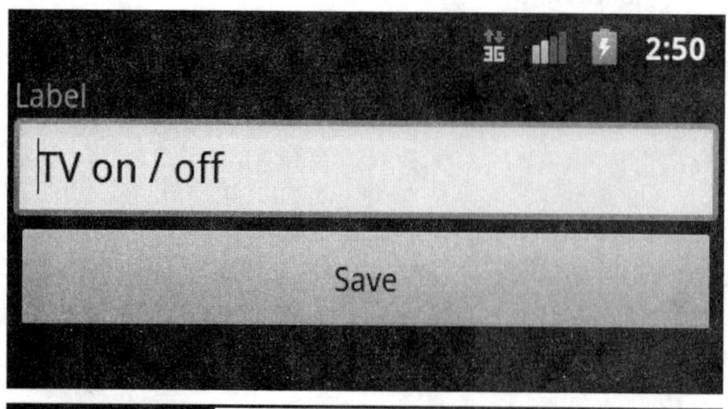

Figure 4-9 Editing a button name

Theory

In this "Theory" section, we will look at how IR remote controls work.

IR Remote Controls

Infrared remote controls send commands as a stream of pulses. Those pulses are modulated at a carrier frequency that is around 40 kHz. Figure 4-10 shows the waveform of a typical message.

The lengths of the codes vary between 12 and 32 bits. A "1" is usually indicated by a "mark" (LED sender on), and a "0" by a space (LED off). The remote will often repeat the message a number of times when a button is pressed.

Figure 4-10 The oscilloscope trace of an IR message

Standards are definitely a concept where less is more. Unfortunately, in the world of IR remotes, there are many standards. This makes it more practical to record the signals you wish to send than to try and synthesize them knowing the make and model of the appliance. There are just too many different variations of code.

Summary

Now that the Evil Genius can control his TV from his phone, his next project is to create another Open Accessory device. This time, a temperature logger that records temperatures and has them wirelessly logged to the Pachube telemetry site.

CHAPTER 5

Temperature Logger

The Evil Genius has a cat (this is to be expected). He is very fond of this cat and is concerned about the animal's well-being. So, he required some reassurance that the lair was warm enough at night for his feline. He therefore decided to build this particular project, which would log the temperature in the cat's suite (Figure 5-1).

The Android control software for the project is shown in Figure 5-2.

An interesting feature of this temperature logger is that it also uses the Android phone to transmit the readings it takes every minute to the Pachube web site, where you can do all sorts of clever charting and analysis (Figure 5-3).

Figure 5-1 The temperature logger Android accessory

Figure 5-2 The Android Temperature Logger app

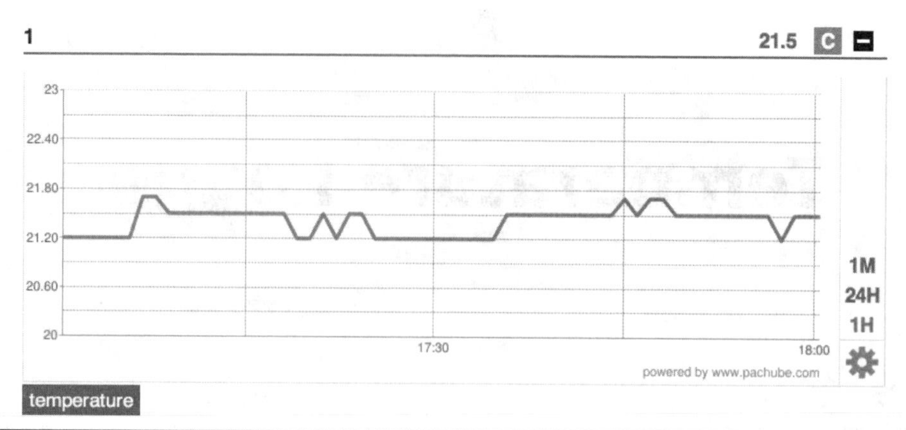

Figure 5-3 The Pachube web site

Construction

Once you have constructed the Droid Accessory Base from Chapter 3, there is really very little construction to do. In fact, there is hardly any soldering to perform at all. The biggest task is probably to put the project in a box.

The schematic diagram for the project is shown in Figure 5-4.

The project uses just three of the IO pins on our Droid Accessory Base. It uses pin D2 as a digital input from a temperature sensor and D6 to supply power to the sensor.

The temperature sensor used in this project is a very interesting device. It combines a temperature sensor and a processor that sends the temperature in digital form. This allows it to be positioned a distance away from the Arduino, without the length of the wire causing problems with the accuracy of the temperature measurement.

What You Will Need

In addition to an accessory-capable Android phone (Android 2.3.4 or later) and all the components required to make the Droid Accessory Base, you just need an ultrasonic range finding–module, a laser module resistor, and a few connectors. See the Parts Bin on the next page.

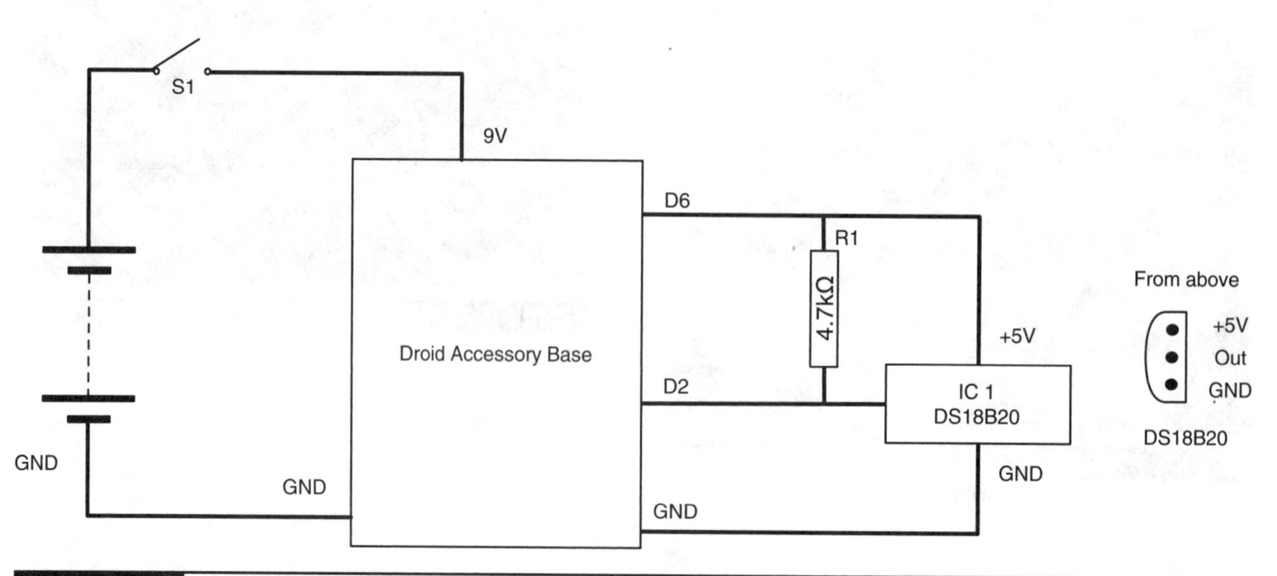

Figure 5-4 The schematic diagram

PARTS BIN

Part	Quantity	Description	Source
Droid Accessory Base	1	*See Chapter 3 for parts and building instructions*	
R1	1	4.7kΩ 0.5W metal film resistor	Farnell: 9340629
R	Optional	1kΩ 0.5W metal film resistor	Farnell: 9339779
IC1	1	DS18B20	SparkFun: SEN-00245
Screw terminal	1	Three-way screw terminal	Farnell: 1641933
Battery clip	1	9V battery clip	Farnell: 1183124
Case	1	Plastic case	Local hardware store

The only electronic components in this project are the temperature sensor chip and a single pull-up resistor.

If you plan to make the noncharging USB lead, you will also need a 1k resistor and an old USB extension lead. See Chapter 2 for details on how to make this lead, which will prevent your mobile phone from draining your accessories' batteries.

As a longer-lasting battery alternative, you can use a holder that will accept six AA cells to supply the project at 9V.

In addition to these components, you will also need the following tools listed in the Toolbox.

TOOLBOX

- Soldering equipment
- An electric drill and assorted drill bits
- Assorted self-tapping screws
- A computer to program the Arduino
- A USB-type A-to-B lead

Step 1. Make a Droid Accessory Base

This is the module described in Chapter 3 that is comprised of a USB host shield, an ATMega328 microprocessor programmed in your Arduino, and a few other components. Follow the instructions in Chapter 3 to construct this.

Step 2. Attach the Components to the Screw Terminal

As we have so few components in this project, the temperature IC and resistor can be attached to a screw terminal (Figure 5-5) that can be plugged

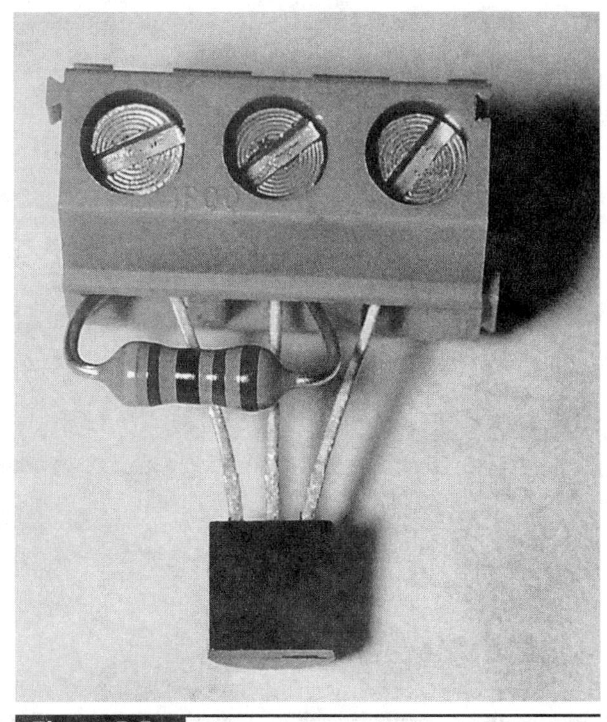

Figure 5-5 The screw terminal

directly into the socket connector on the Droid Accessory Base (Figure 5-6). Make sure the small three-pin IC is the correct way around. The flat side of the chip should be facing upward.

Step 3. Install the Arduino Sketch

To get the sketch onto the microcontroller, we are first going to program the chip and then carefully take it off the Arduino board and fit it onto the socket on the Droid Accessory Base.

This sketch makes use of two libraries for the temperature module (OneWire and DallasTemperature).

The procedure for installing a library is the same as for all libraries (see Chapter 1). You can download the OneWire and Dallas libraries in a single zip file from www.hacktronics.com/code/OneWire-v2.zip.

So, just install the sketch (ch05_temp_logger) onto the Arduino board. Then, with the power disconnected, carefully remove the IC from the board and fit it into the socket on the Droid Duino Base, making sure it's the right way around. The easiest way to remove the IC is to gently lever it up from each end in turn, a little at a time, using a small screwdriver.

Step 4. Install the Android App

If your phone has an Internet connection, you can skip installing the Android app and wait until the phone's Open Accessory software asks for it, which will be when you first connect the range-finder accessory. Otherwise, visit www.duinodroid.com and download the APK installer.

If you have any trouble with this, read through Chapter 1, Step 10.

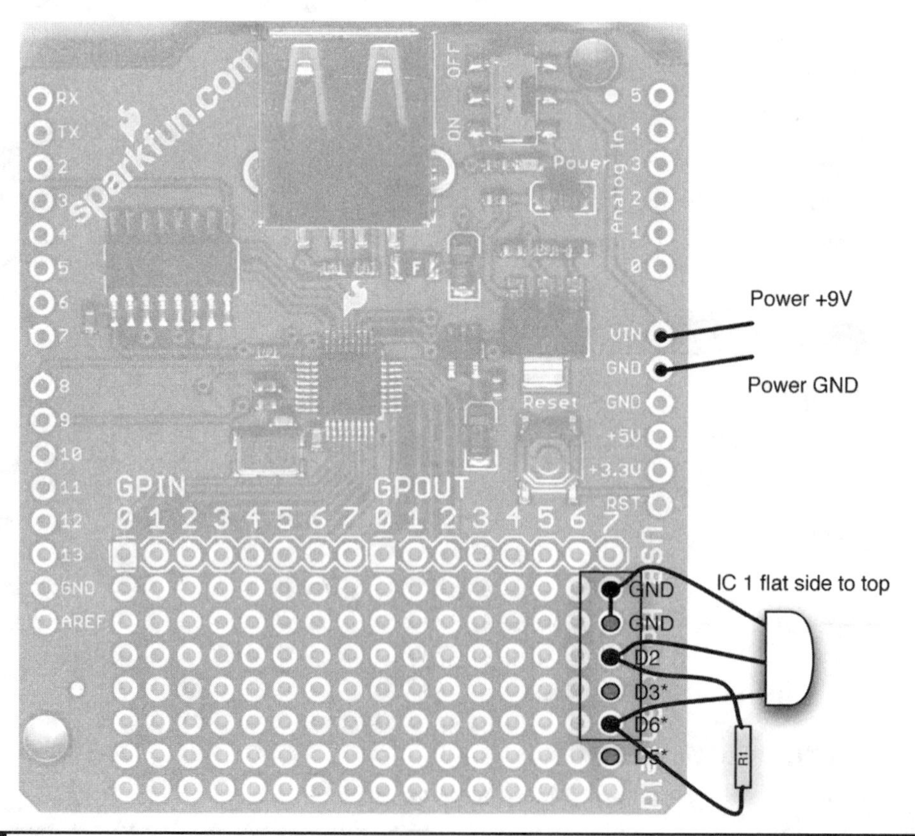

Figure 5-6 The wiring diagram

Step 5. Test

For this project, the Droid Accessory Base has a 9V battery clip soldered to the power leads.

It is always a good idea to test your project before you fit everything into a box. So, connect the battery and your phone (Figure 5-7).

Turn on the switch and it should cause the phone to either launch the Droid Temp Logger app, or if it is not installed, to display the URL from where it can be downloaded.

Step 6. Box the Project

Boxing this project is a very similar process to the three previous projects. First, lay out the components in the box (Figure 5-8), and then work out where you will need holes and how big they should be.

You will need holes for:

- The USB cable
- Screws in the base of the box to hold the board in place
- The switch
- A hole near the temperature sensor

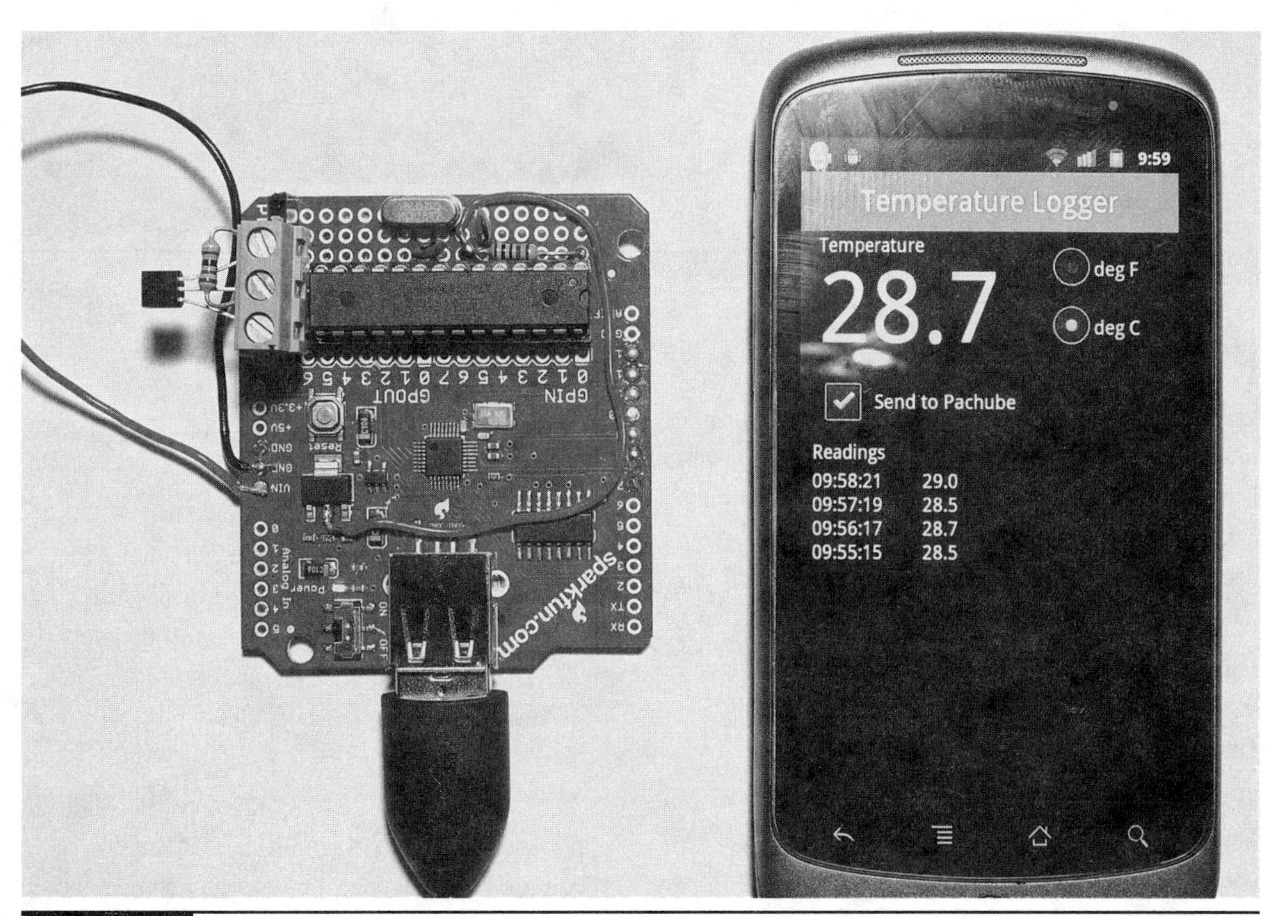

Figure 5-7 Testing

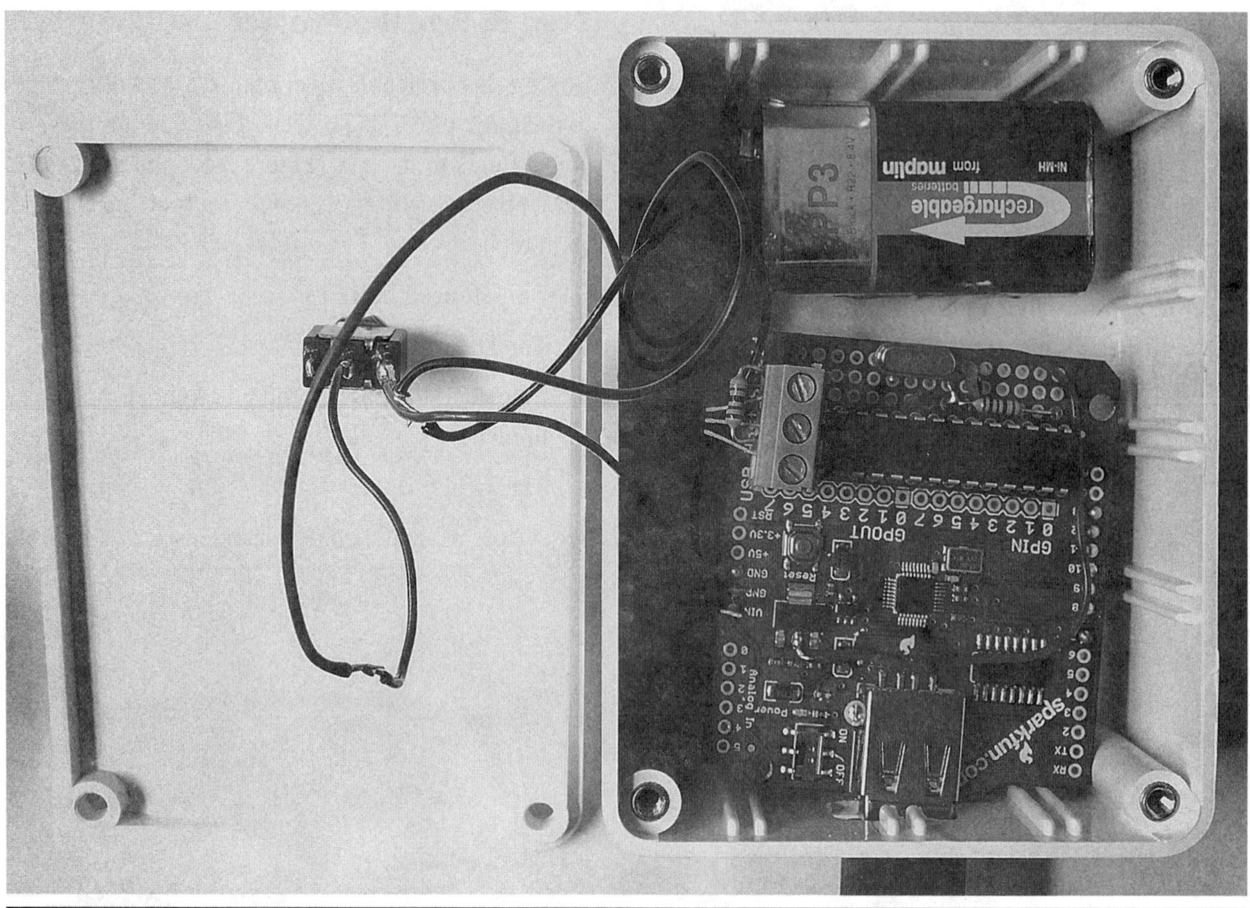

Figure 5-8 Laying out the components in a box

Using the Project

To really get the most out of this project, we need to set up an account with Pachube. This will allow us to automatically send temperature readings that arrive on the Android phone out to the Pachube account.

Start by creating a new Pachube account for yourself on the web site www.parchube.com. This is free for Pachube Basic members, which will be fine for our purposes.

Once your account has been created, you will have to click "Create a Feed" to give our data somewhere to go. Figure 5-9 shows this. The only mandatory field is the name for the feed. You also need to create a stream, so click the add stream and enter the details shown in Figure 5-10. Afterward, click Save. You will then see a summary of your new feed.

Make a note of the number—in this case, 29933. You will need to enter this into your Android application. The other piece of information you will need is your personal API key for your account. You can find this by clicking My API Keys on the right-hand side of the web site. This will be a long key, something like: B74W43hXupkN6J_rEsM1exa30X6TjP8cNFFDY s2i6r.

You will also need to enter this into the Android app, but it is easier to log in to the Pachube account on your phone's browser so you can copy and paste it.

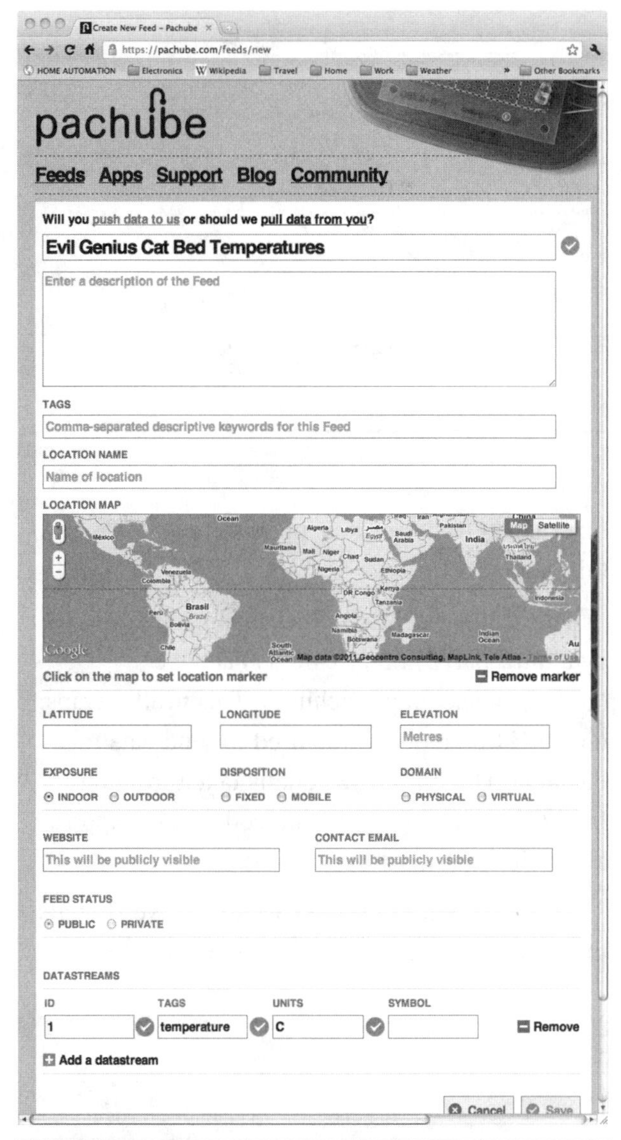

Figure 5-9 Creating a stream in Pachube

Now that we have everything ready to go, plug in the accessory to launch the TempLogger app, click the Menu button on your phone, and select Settings.

Enter the Feed ID and Key as shown in Figure 5-11.

Once you have entered the settings, click Save and then use the phone's Back button to return to the main screen.

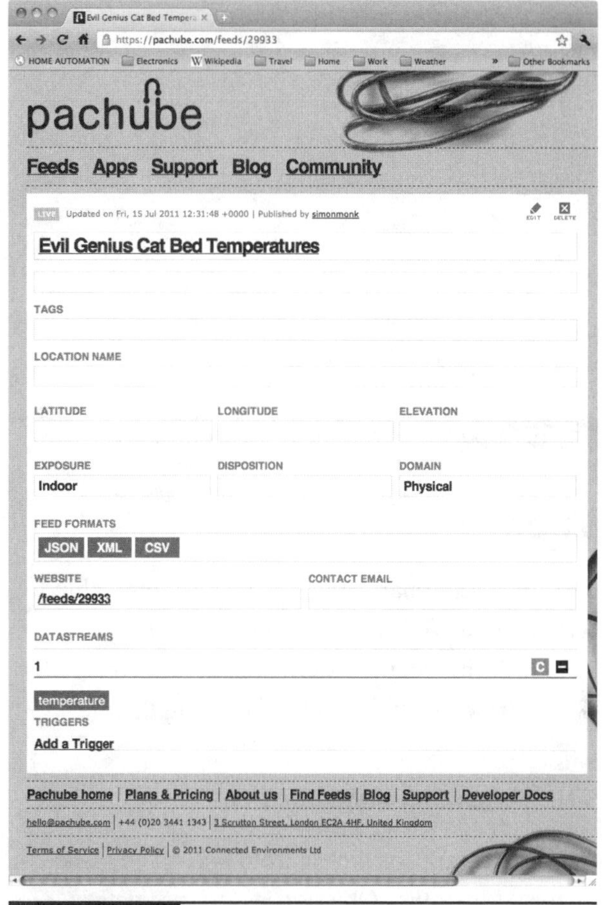

Figure 5-10 The feed summary

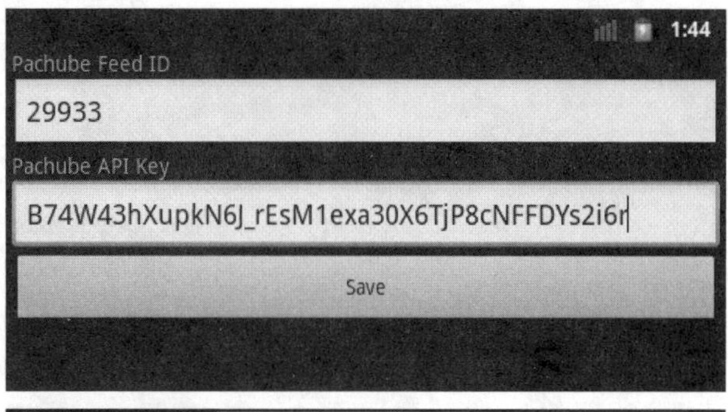

Figure 5-11 Setting the Feed ID and Key for Pachube

Click the checkbox to log in to Pachube. Leave the logger running for a while. You can then go and see your data on the Pachube web site (Figure 5-12).

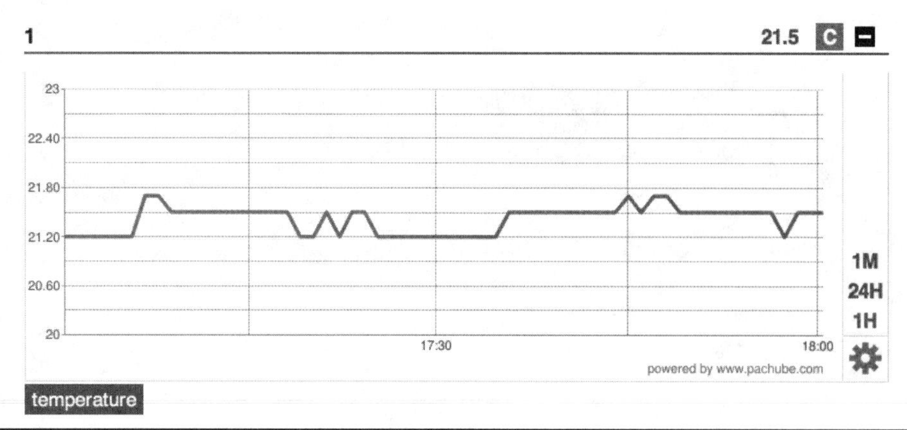

Figure 5-12 The temperature chart

Theory

The interesting part of this project is being able to wirelessly record remote temperatures to a web server. In this section, we will look at the part of the Android app responsible for doing this.

The Android Application

The framework for the Android app is the same as for all the other Open Accessory projects in this book, so please have a look at one of the other Open Accessory chapters for details on how this works.

The sending of Pachube data is handled by a third-party library called JPachube (code.google.com/p/jpachube). This neatly wraps up the HTTP requests required to send data to Pachube. The following code is taken from the InputController class. All the code for the app is available from www.duinodroid.com.

The code is fairly self-explanatory. We first get the Feed ID and Pachube Key out of the

```
private void sendPachubeReading() {
  SharedPreferences settings =
    mHostActivity.getSharedPreferences(SettingsActivity.PREFS_NAME, 0);
  String pachubeFeed = settings.getString("PACHUBEFEEDID", null);
  String pachubeKey = settings.getString("PACHUBEKEY", null);
  if (pachubeFeed == null || pachubeKey == null) {
    alert("Go to Settings and add entries for Pachube Feed ID and Key.");
  }
  else {
    try {
      Pachube p = new Pachube(pachubeKey);
      Feed f = p.getFeed(Integer.parseInt(pachubeFeed));
      Double reading = Double.parseDouble(mTemperature.getText().toString());
      f.updateDatastream(1, reading);
      }
      catch (PachubeException e) {
        alert(e.errorMessage);
      }
    }
}
```

preferences system. If they do not have a value, then the user is prompted to go to the settings page and make the necessary changes.

It's then simply a matter of using the Pachube and Feed classes to send the data. In the event of an error, the error message is reported to the user in an alert.

Summary

We now have another Open Accessory project under our belt. In the next chapter, we will move from measuring temperature to measuring distance—using an ultrasonic range finder.

CHAPTER 6

Ultrasonic Range Finder

AS YOU MIGHT EXPECT, the Evil Genius has many friends in the real estate business. One of these friends mentioned that it was extremely hard work measuring up rooms using a traditional tape measure while clutching their cell phone. Apparently, many realtors are afflicted by a medical condition that makes it impossible to take their cell phone from their hand.

This inspired the Evil Genius to design the final project in his Android accessory lineup, an ultrasonic range finder (Figure 6-1).

The Android control software for the project is shown in Figure 6-2.

Figure 6-1 The Range Finder Android accessory

Figure 6-2 The Android Range Finder app

Because almost any project is improved by the addition of a laser, the Evil Genius decided to include a laser on the accessory to indicate the direction in which the distance is being measured.

CAUTION You should take certain precautions when using lasers: (1) Never shine a laser into your, or anyone else's, eyes. (2) Resist the temptation to check if it's on by peering into it. Instead, always shine it onto paper or some other light colored surface.

Construction

If you have already constructed the Droid Accessory Base from Chapter 3, there is really very little construction to do here. In fact, there is hardly any soldering to perform—the biggest task is probably just putting the project in a box.

The schematic diagram for the project is shown in Figure 6-3.

The project uses all four of the IO pins on our Droid Accessory Base. It uses one pin to provide a +5V supply to the ultrasonic module, two pins for the "trigger" and "echo," and the final pin to control the power to the laser.

What You Will Need

In addition to an accessory-capable Android phone (Android 2.3.4 or later) and all the components required to make the "Droid Accessory Base," you just need an ultrasonic range–finding module, a laser module, a resistor, and a few connectors. See the Parts Bin for a full list of what you need.

The key component in this project is the ultrasonic range–finder module. Figure 6-4 shows the module used by the author. This module requires just four pins to operate it. As well as GND (ground or 0V) and 5V, the module has a "trigger" input and an "echo" output. The datasheet for this component specifies that you must provide a 10-microsecond

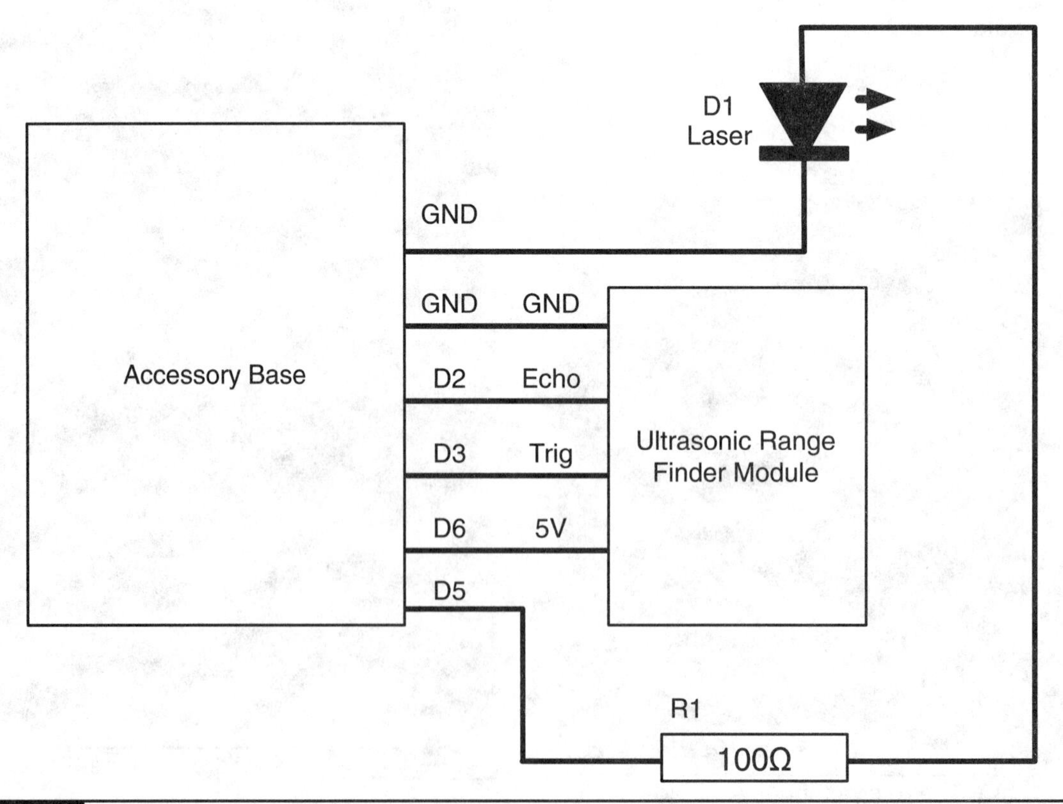

Figure 6-3 The schematic diagram

PARTS BIN			
Part	**Quantity**	**Description**	**Source**
Droid Accessory Base	1	*See Chapter 3 for parts and build instructions*	
R1	1	100Ω 0.5-W metal film resistor	Farnell: 9339760
Battery	1	PP3 9V NiMH rechargeable battery	
Battery clip	1	PP3-style battery clip	Farnell: 1183124
D1	1	Laser diode module 5 mW	eBay
Case	1	Plastic case 100 × 76 × 41mm	Local electronics store

Figure 6-4 The ultrasonic range-finder module

pulse to the trigger input and then see how much time elapses before the "echo" pin indicates that the sound wave has bounced back. See the "Theory" section later in this chapter for more details on how ultrasonic range finding works.

It will make construction much easier if you can find a module similar to that used by the author. The interfaces to these devices all seem much the same, and if you can find one with the pins in the same places, that will be a big help. The author sourced his module from eBay, where there were many similar devices on offer for just a few dollars.

If you plan to make a noncharging USB lead, you will need two 1kΩ resistors rather than just one. You will also need an old USB extension lead. See Chapter 2 for details on how to make this lead, which will prevent your mobile phone from draining your accessories' batteries.

As a longer-lasting battery alternative, you can use a holder that will accept six AA cells to supply the project at 9V.

In addition to these components, you will also need the following equipment shown in the Toolbox.

TOOLBOX

- An electric drill and assorted drill bits
- Assorted self-tapping screws
- A computer to program the Arduino
- A soldering iron and solder
- An A-to-B lead

Step 1. Make a Droid Accessory Base

This is the unit described in Chapter 3 that comprises a USB host shield, an ATMega328 microprocessor programmed in your Arduino, and a few other components.

Step 2. Solder the Laser and Resistor

The laser module will have two leads: a red positive connection and a black negative connection. You will probably need to shorten the leads to a couple of inches (50mm). Strip and tin the ends of the leads, putting a fairly thick layer of solder on the negative lead, which we are going to push straight into the top-most header pin on our Droid Accessory Base. Shorten the leads of the 100Ω-resistor R1 and solder one end to the positive laser lead. The other end of the resistor is also going to be pushed into the socket for the D5 connection, so thicken this up a little with solder.

The completed laser module is shown in Figure 6-5.

Step 3. Install the Arduino Sketch

To get the sketch onto the microcontroller, we are first going to program the chip and then carefully take it off the Arduino board and fit it onto the socket on the Droid Duino Base.

The sketch uses the same libraries as the projects in Chapters 2 thru 5, so if you have not completed one of these projects, then refer to the "Construction" section in Chapter 2, under the subsection "Step 6. Install the Open Accessory Libraries."

Once the sketch is installed on the microcontroller, carefully remove it, levering each end up a little at a time so as not to bend the pins. Then, seat it onto the Droid Accessory Base, making sure you have it the right way around.

Step 4. Install the Android App

If your phone has an Internet connection, you can skip installation of the Android app and wait until the phone's Open Accessory software asks for it (when you first connect the range finder accessory). Otherwise, visit www.duinodroid.com and download the APK installer.

If you have any trouble with this, return to Chapter 2 and read "Step 10" in the "Construction" sequence section.

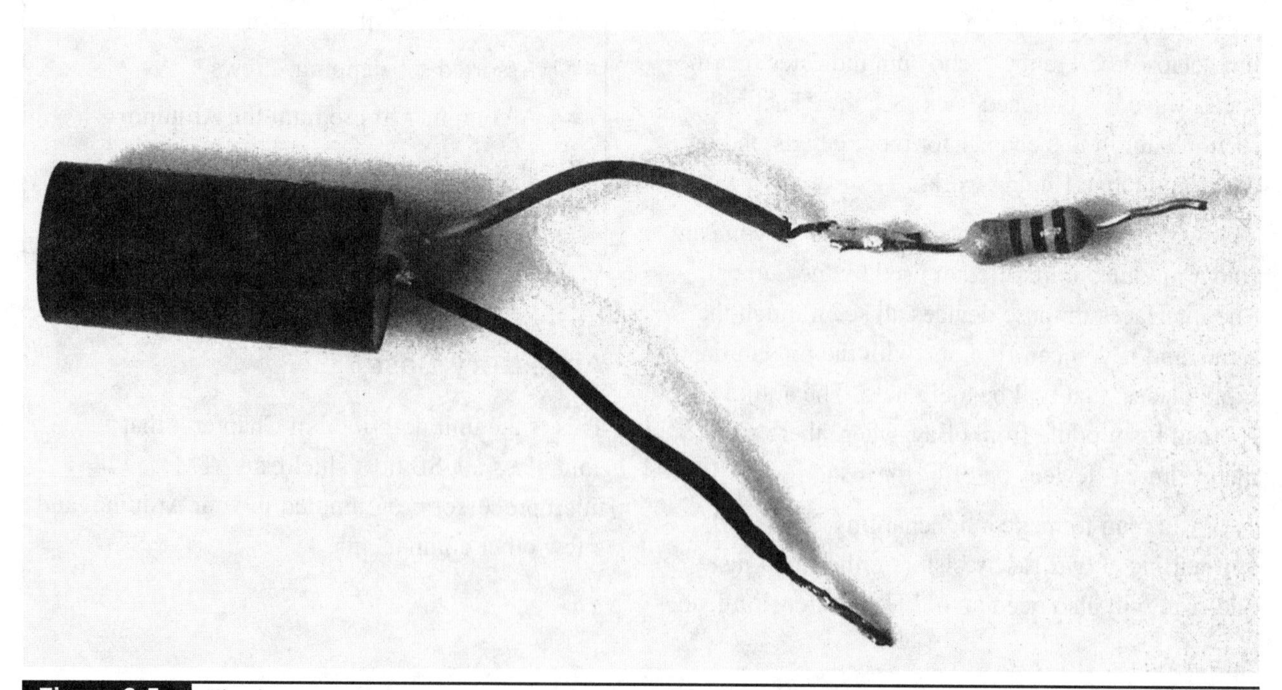

Figure 6-5 The laser module

Step 5. Test the Project

It is always a good idea to test your project before fitting everything into a box. So, connect everything up, as shown in Figure 6-6.

Now, turn on the switch. This should cause the phone to either launch the Droid Range Finder app or, if it is not installed, to display the URL from where it can be downloaded.

Try clicking the Laser checkbox to make sure it turns the laser on and off. Then, experiment by moving your hand in front of the ultrasonic transducers and notice the distance reading changing in the Android display screen. Pressing the current reading will record it to the log.

Step 6. Box the Project

Lay the components out in the box, working out where everything will go, and mark the positions for the hole with a pencil (Figure 6-7).

Drill the holes for the ultrasonic transducers, the laser module, the USB socket and two small holes for fixing the shield board to the bottom of the box. You will also need a hole in the side of the box for the switch.

Use small self-tapping screws to fix the shield into the box (Figure 6-8). These should be small enough to hold the board in place without sticking out of the bottom of the box, where they could scratch something.

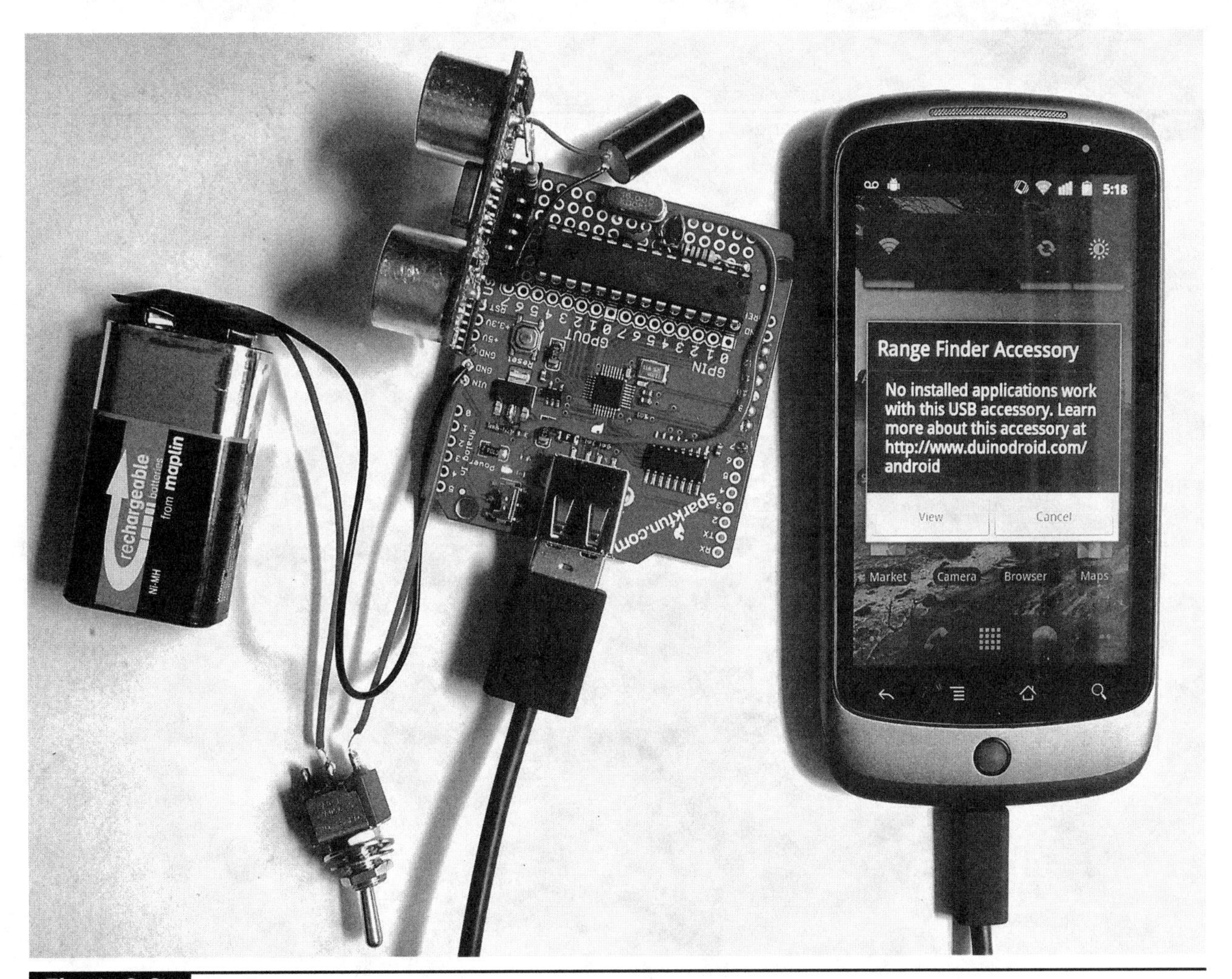

Figure 6-6 Testing

Figure 6-7 Positioning the components in the box

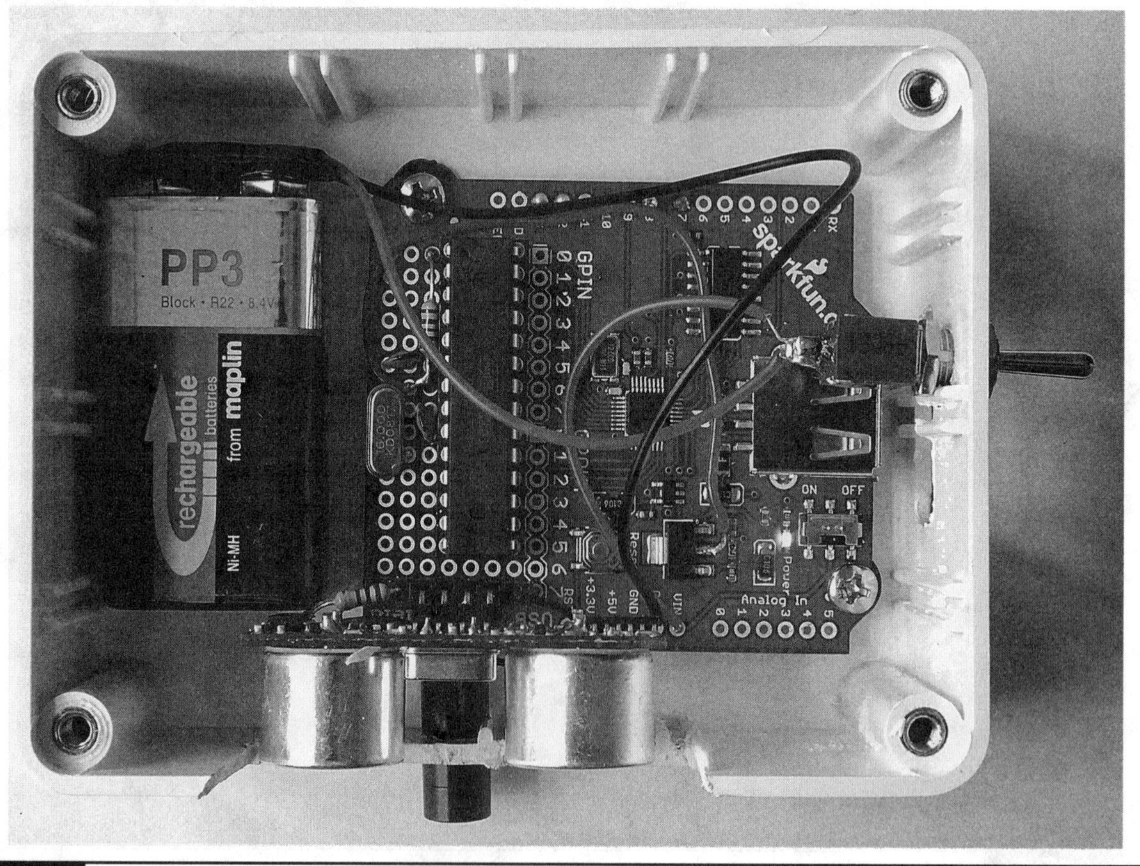

Figure 6-8 The components fixed inside the box

Using the Project

By now, you should have the Android app working and continuously displaying the distance to the target (Figure 6-9).

You can change the units between inches and cm using the radio buttons, and by tapping the reading itself at any time, the reading will be added to the list of readings at the bottom of the screen.

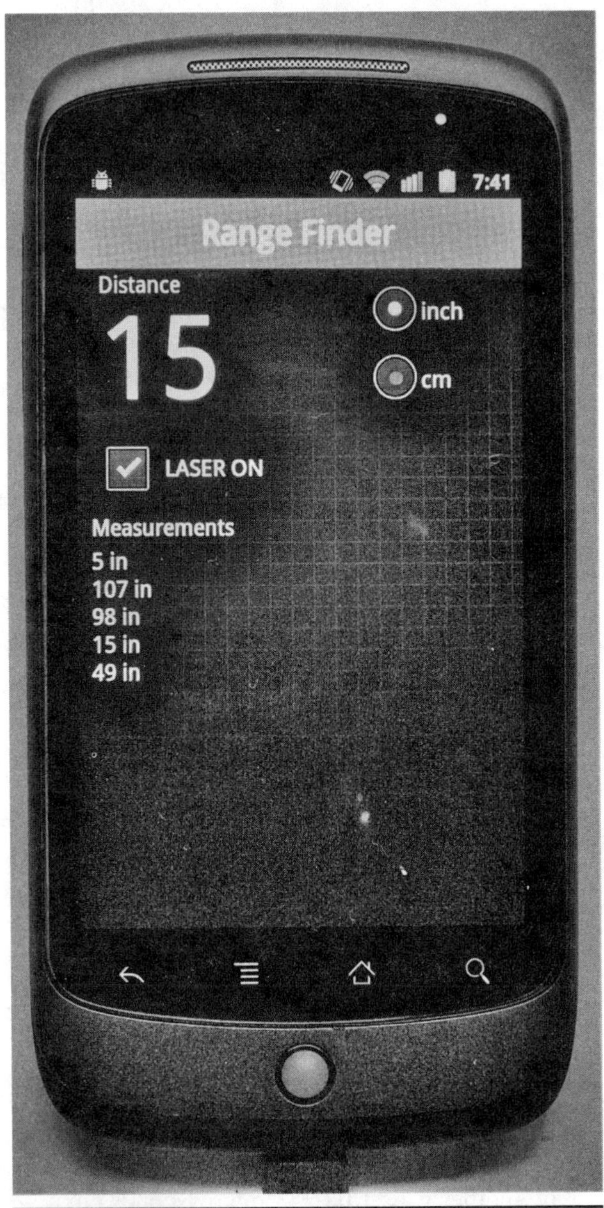

Figure 6-9 The Android app

The checkbox allows the laser to be turned on and off.

The source code for the app is available as open-source software; see the book's web site (www.duinodroid.com). Please feel free to experiment with and improve the code, because you can take this project in many interesting directions, including:

- Automatically detecting that something has moved and sending an SMS or web request
- Calculating the area of a room from the two dimensions

Theory

We have now looked at enough Android accessories in earlier chapters to not really wish to repeat ourselves here.

We will, however, look at how ultrasonic range finding works, and how you interface to this in an Arduino sketch.

Ultrasonic Range Finding

Ultrasonic range finding works the same way as sonar used by ships and submarines. A sound wave is sent out from a sender, hits an object, and bounces back. As you know, using the speed of sound, the distance to the sound-reflecting object can be calculated by measuring the time it takes for the sound pulse to return to the receiver (Figure 6-10).

The sound used is at a high frequency—hence, it is called *ultrasonic*. Most units operate at a frequency of about 40 kHz. Not many people can hear sounds above 20 kHz.

The Arduino code for measuring the range is all contained within the takeSounding function. This sends a single 10-microsecond pulse to the "trigger" pin of the ultrasonic module, which then uses the built-in pulseIn Arduino function

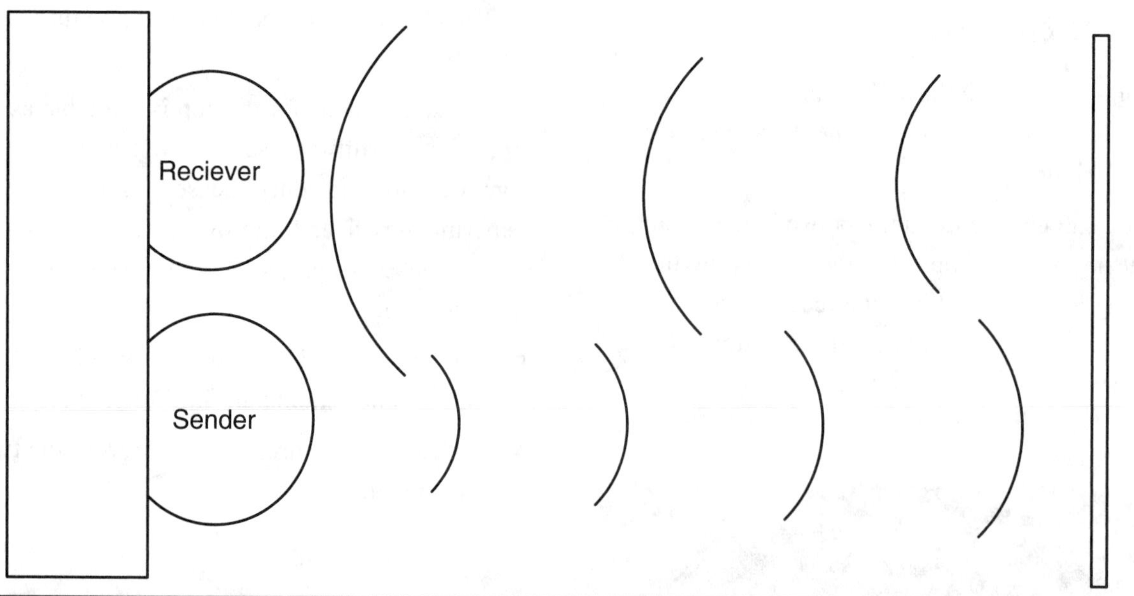

Figure 6-10 Ultrasonic range finding

to measure the time period before the echo pin goes high.

This period is the amount of time it took for the sound wave to travel from the sender, hit the target, and bounce back to the sensor.

```
long takeSounding()
{
  digitalWrite(trigPin, LOW);
  delayMicroseconds(2);
  digitalWrite(trigPin, HIGH);
  delayMicroseconds(10);
  digitalWrite(trigPin, LOW);
  delayMicroseconds(2);
  long duration = pulseIn(echoPin,
   HIGH);
  long distance =
   microsecondsToCentimeters(duration);
  if (distance > 500)
  {
    return lastDistance;
  }
  else
  {
    lastDistance = distance;
    return distance;
  }
}
```

We then need to convert that time in milliseconds into a distance in centimeters. If there is no reflection, because there is no object that is close enough, or the object is reflecting the sound wave away, rather than bouncing it back directly to the receiver, the time of the pulse will be very large, and so the distance will appear as very large.

To filter out these long readings, we disregard any measurement that is greater than 5m, using the last sensible reading we got.

The function microsecondsToCentimeters is actually very straightforward.

```
long microsecondsToCentimeters(long
     microseconds)
{
  return microseconds / 29 / 2;
}
```

The speed of sound is roughly 343m/s in dry air at 20 °C, or 34,300cm/s.

Or, put another way, 34,300/1,000,000cm/microsecond.

That is, 0.0343cm/microsecond.

Put another way, 1/0.0343 microseconds/cm.

Or, 29.15 microseconds/cm.

Thus, a time of 291.5 microseconds would indicate a distance of 10cm.

The function microsecondsToCentimeters rounds off 29.15 to 29 and then also divides the answer by 2, because we don't want the distance of the whole return journey, just the distance to the subject.

In actual fact, many factors affect the speed of sound, so this approach can only give an approximate answer. The temperature and the humidity of the air will both affect the measurement.

Summary

This is the last of our Android accessories. The remainder of the book is devoted to combining Android and Arduino in a different way, to provide a home automation system based on a low-cost Android tablet. Because such devices are not yet ready to use the Android Open Accessory protocol, we will have to find our own way of connecting them to an Arduino.

PART TWO

Home Automation

CHAPTER 7

Home Automation Controller

THIS PROJECT FORMS the hub of the Evil Genius' home automation system. It is the main control system for many of the projects in this book.

Lair Automation is a big deal for the Evil Genius, who likes to remotely control the lights, power sockets, and heating, and even control access to the front door from wherever he is in the world. So, in the event of unexpected visits from out-of-town Evil Genius friends, he can turn up the heating and unlock the door, without having to leave the comfort of his couch. He finds this so much easier than barking orders at his minions.

The Evil Genius does try to be green and generate his own electricity, but the minions can only run around a treadmill for so long before they become tired and hungry. So the lair is still connected to the electric company. The Evil Genius is also somewhat preoccupied by the size of his electricity bill, and despite threats and horrific punishments, his minions will tend to pointlessly leave the lights on in the shark tank or leave the death-ray on standby rather than turn it off at the outlet. The home (or lair) automation controller provides technology to help the Evil Genius turn things off automatically and keep his energy bills under control.

Figure 7-1 shows the control unit for the home automation controller, and Figure 7-2 shows how the home automation controller fits with some of the other projects in this book.

Looking at Figure 7-2, you can see that the controller is based on an Android tablet and an Arduino board linked closely together.

Arduino boards are great for connecting electronics to computers, but are not so good at providing a nice, easy-to-use interface, over and above a small display and a few buttons. This is where the cheap $100 Android tablet comes in. This has a large touchscreen, WiFi, sound output, and sometimes even a webcam. What's more, an Android tablet can act as a web server, so we can access our home automation from any Web device, whether at home or at large via the Internet.

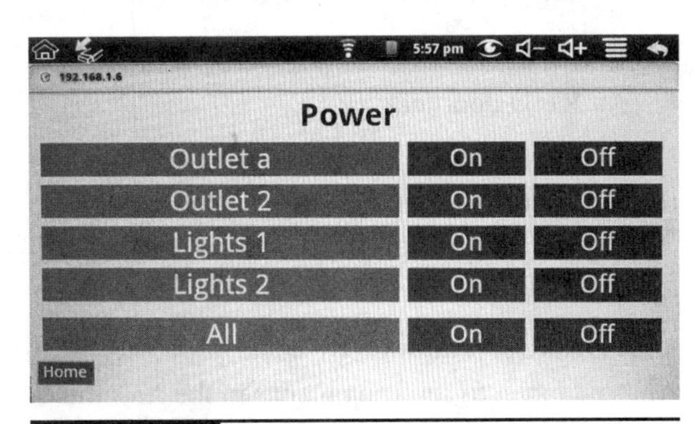

Figure 7-1 The home automation controller

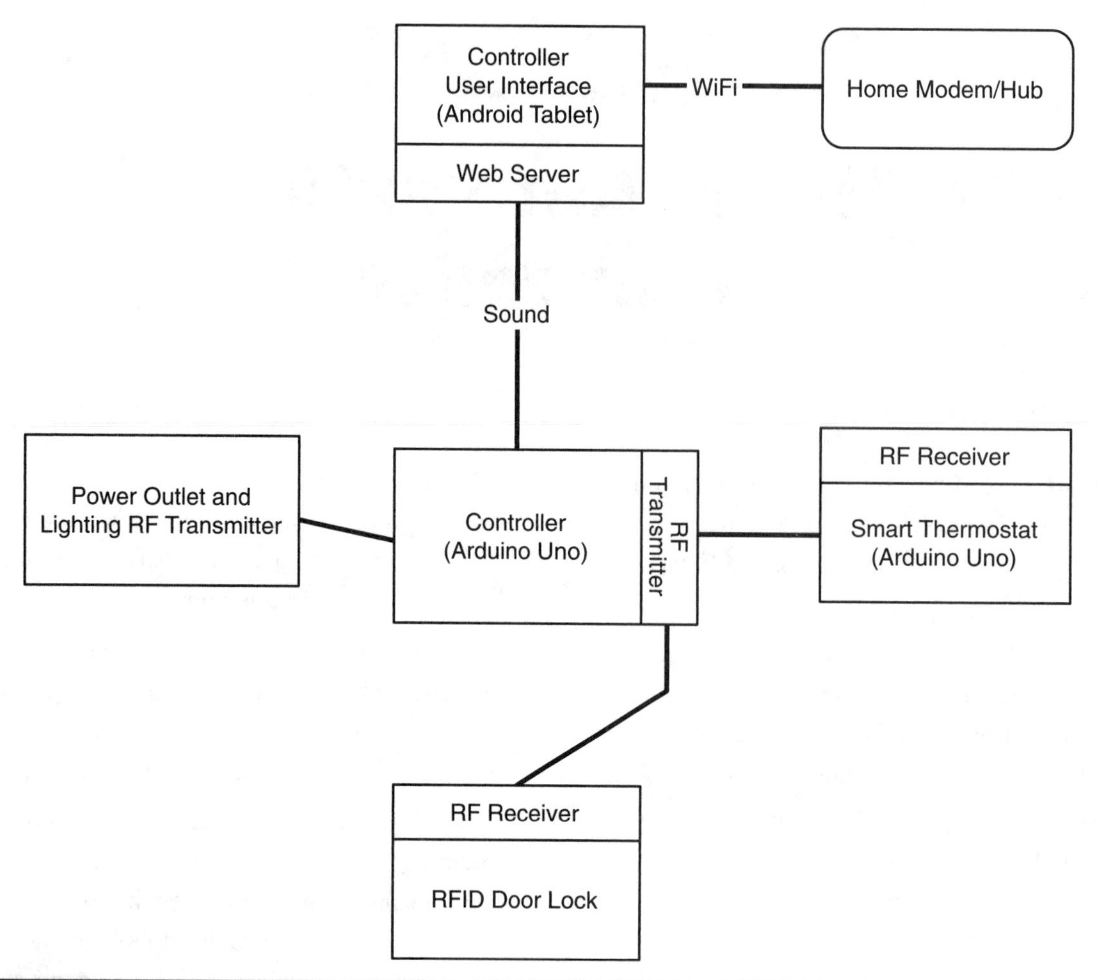

Figure 7-2 The home automation system

Eventually our home automation controller will be linked to:

- RF-controlled wall outlets and lights
- A "smart" thermostat using another Arduino
- An RFID door lock

For now, we will content ourselves with creating the basic setup for the Arduino and Android tablet. As the book progresses, we will add other components to the project a chapter at a time.

There are several ways we could allow the Android tablet to send commands to the Arduino. We could:

- Equip our Arduino with an Ethernet or WiFi shield and communicate over a network.
- Connect a Bluetooth module to the Arduino, and, if our Android tablet has Bluetooth, send data that way.
- Use the headphone output from the Android tablet to send commands to the Arduino.
- Use a USB connection and the Google ADK (this requires Android 2.3.4 or later).

In other chapters, we make use of both Ethernet and Bluetooth, as well as an RF radio link and USB using the Android ADK. But for this project, we are going to build a "sound" interface between the Android and the Arduino. This has the advantage of being a "hard" link—in other words, a wire that runs directly from the Android tablet to the Arduino. Also, many of the low-cost Android tablets do not have Bluetooth or Android 2.3.4 (necessary for ADK), which would otherwise make ADK a good choice.

The Android tablet will generate warbling tones, employing the same technique that early home computers used to use when saving programs onto cassette tape.

The usefulness of this kind of link is not limited to home automation. It can be used in any situation where you need an Android tablet to send commands to an Arduino.

This project is split into three sections: making the sound link hardware; programming the Android tablet; and programming the Arduino.

The Sound Link Module

The schematic diagram for the sound link module is shown in Figure 7-3.

The "Theory" section at the end of this chapter holds more information on how this works, but for now we will content ourselves with knowing that the Android device will produce a sequence of 16 sound pulses at 1 kHz. A long pulse will indicate a logical 1, and a short pulse a 0. These pulses are detected and fed into a digital input on the Arduino, which converts them to a number.

What You Will Need

You will need the following components listed in the Parts Bin to make the sound link module.

PARTS BIN

Part	Quantity	Description	Source
Arduino Uno	1	Arduino Uno board	www.arduino.cc
Android tablet	1	Low-cost Android tablet—Android version 2.1 or later, with headphone jack and WiFi	eBay
15V PSU	1	15V 1A wall-wart power supply	Farnell: 1176620
R1, R3	2	1kΩ 0.5-W metal film resistor	Farnell: 9339779
R2	1	1MΩ 0.5-W metal film resistor	Farnell: 9339809
R4	1	100kΩ 0.5-W metal film resistor	Farnell: 9339795
R5	1	33kΩ 0.5-W metal film resistor	Farnell: 9340424
C1	1	1μF 16V electrolytic capacitor	Farnell: 1236655
C2	1	220nF ceramic capacitor	Farnell: 1216441
C3, C4	2	100μF 16V electrolytic capacitor	Farnell: 1136275
C5	1	100nF ceramic capacitor	Farnell: 1200414
D1	1	1N4001	Farnell: 1458986
IC1	1	7611 CMOS operational amplifier	Farnell: 1018166
IC2	1	7909 linear voltage regulator	Farnell: 7202164
IC socket	Optional	Eight-pin DIL IC socket	Farnell: 1101345
	1	Stripboard, 17 rows of 12 holes	Farnell: 1201473
	1	2.1mm power plug	Farnell: 1200147

(continued)

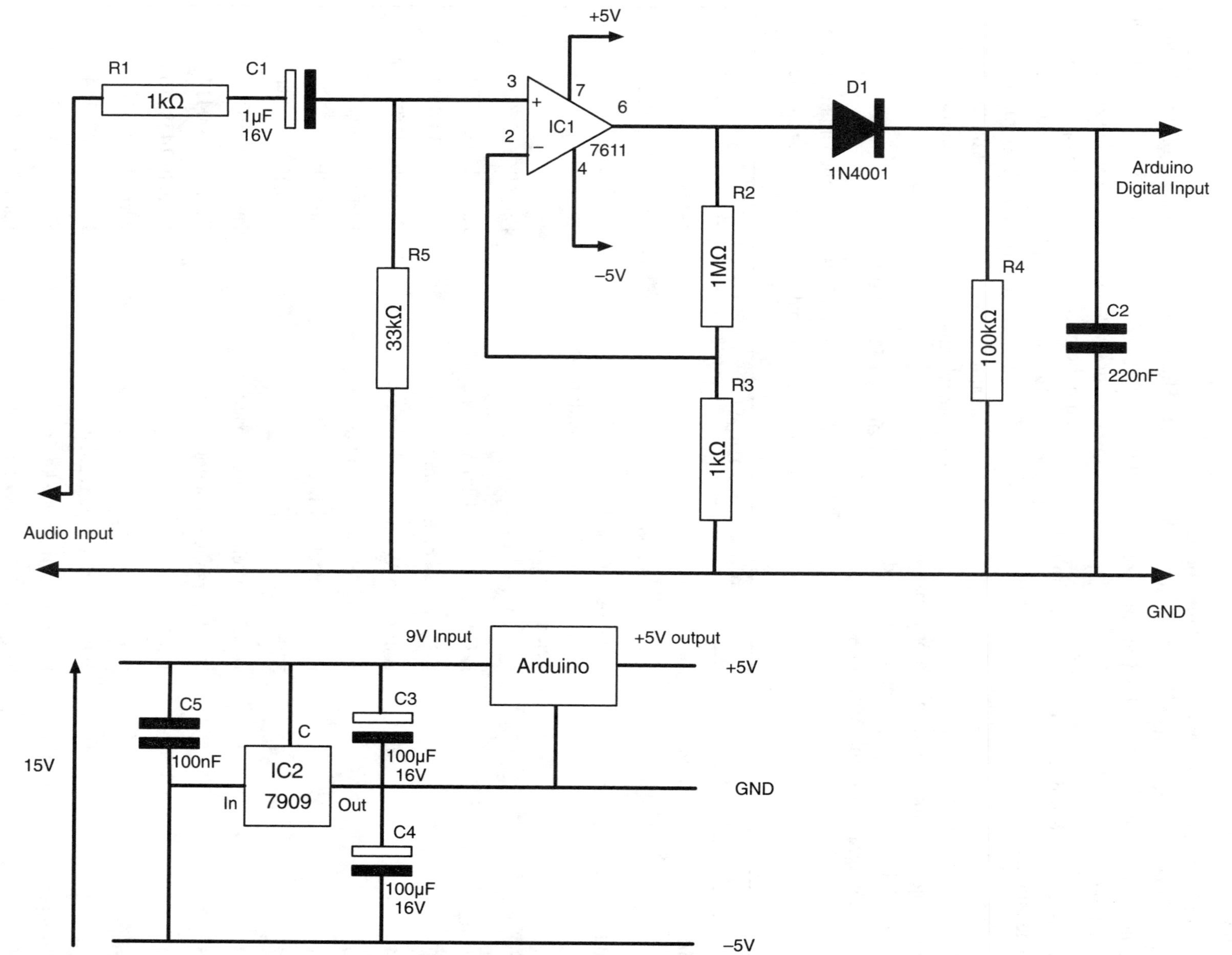

Figure 7-3 The schematic diagram for the sound link

PARTS BIN *(continued)*

Part	Quantity	Description	Source
	1	2.1mm power socket	Farnell: 1217038
3.5mm plug	1	3.5mm stereo audio plug	Farnell: 1267389
Wire		Assorted solid core and multi-core wire *(see the following paragraphs in this section)*	Local electronics store
Box		ABS project box to fit the Android tablet	Farnell, Radio Shack, and others
Strip of steel			Hobby store

Obviously, the most expensive parts of this project are the Android tablet and the Arduino board. The Android tablet that the author used cost around $100 on eBay. It has a seven-inch display and uses Android 2.1. It is clearly designed as a low-cost cure for iPad-envy. The build quality is fairly low, but at that price, what do you expect?

The tablet needs to have Android version 2.1 or later installed on it. It must also have a headphone jack and WiFi, but Bluetooth is not required.

The latest Arduino board is the Arduino Uno. The official Arduino web site (www.arduino.cc) lists suppliers of the Uno. However, if you are on a budget, you can use an Arduino clone of either the Arduino Uno or the older Arduino Duemilanove, which will work just fine.

You can buy a new audio plug, but you may find that you have an old pair of unwanted earphones from which you can just cut off the plug and ten inches (250mm) of lead. This has the added advantage that it already has a lead soldered to it. The same argument applies to the 2.1mm power plug. You may well have an unwanted power adapter from which a plug and short length of lead can be salvaged.

If, like the author, the project box you are going to fit everything into does not have a lot of spare room, then use right-angle plugs, which will stick out from the side of the Android tablet much less. You can also bend the leads on a regular 3.5mm jack plug to stop it from sticking out too much.

As you can see from Figure 7-1, the Android tablet is going to be fixed to the inside of a plastic project box. This needs to be big enough to contain the tablet, but also have enough depth to contain the other electronics that need to be in the box. These components will include the Arduino, the sound interface and later on an RF remote control, a Bluetooth module, and an RF transmitter. This means you will need at least 1 inch (25mm) of depth more than the depth of the tablet. The author's box has a total depth of about 2½ inches (65mm).

For all the projects in this book, you will need a basic stock of wire. For making links on stripboard, solid core wire is best. Something like 22 AWG copper wire is fine. When it comes to making connections between components of the design, such as between the sound interface stripboard and the Arduino board, it's best to use multi-core "equipment wire." Again, around 22 AWG is fine. The author keeps a box of wires reclaimed from old household electronics. It is always useful to have a variety of colors, and to use red for positive connections, black or blue for ground and negative connections, and another color such as yellow or orange for signal connections.

In addition to these components, you will need the following tools listed in the Toolbox.

TOOLBOX
■ Soldering equipment
■ Multi-core wire in various colors
■ A multimeter
■ An electric drill and assorted drill bits
■ A craft knife or strong scissors
■ An optional rotary cutter such as a Dremel

Step 1. Prepare the Stripboard

The sound link is constructed on a small piece of stripboard (Figure 7-4).

If this is the first time you have used stripboard, then there are a few things you need to know.

Stripboard is a perforated board with holes at 1/10-inch pitch. Behind the holes are strips of copper. Component leads are pushed through from the plain side, and soldered to the copper strips. Figure 7-5 shows the stripboard for the sound link from the copper side.

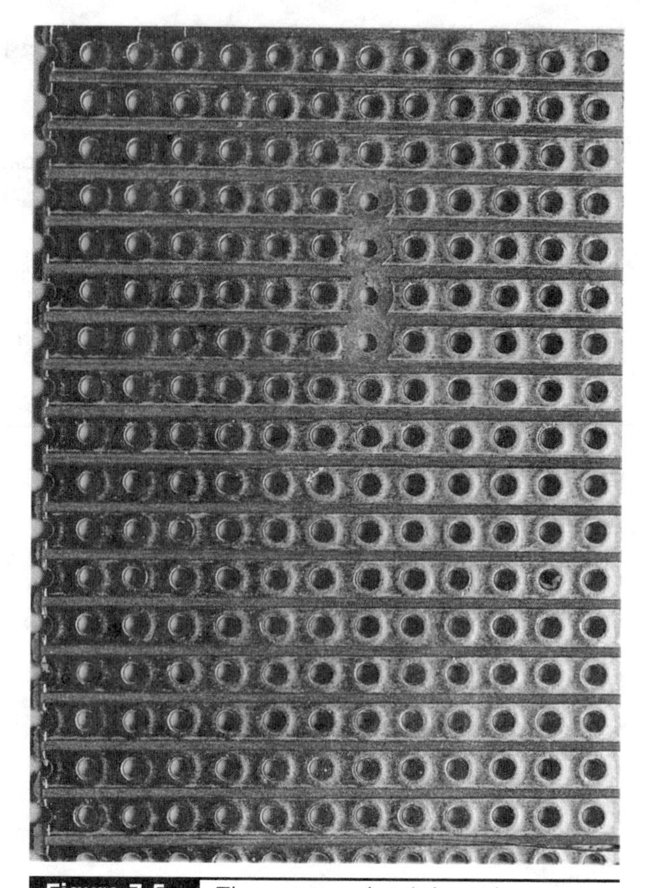

Figure 7-5 The prepared stripboard

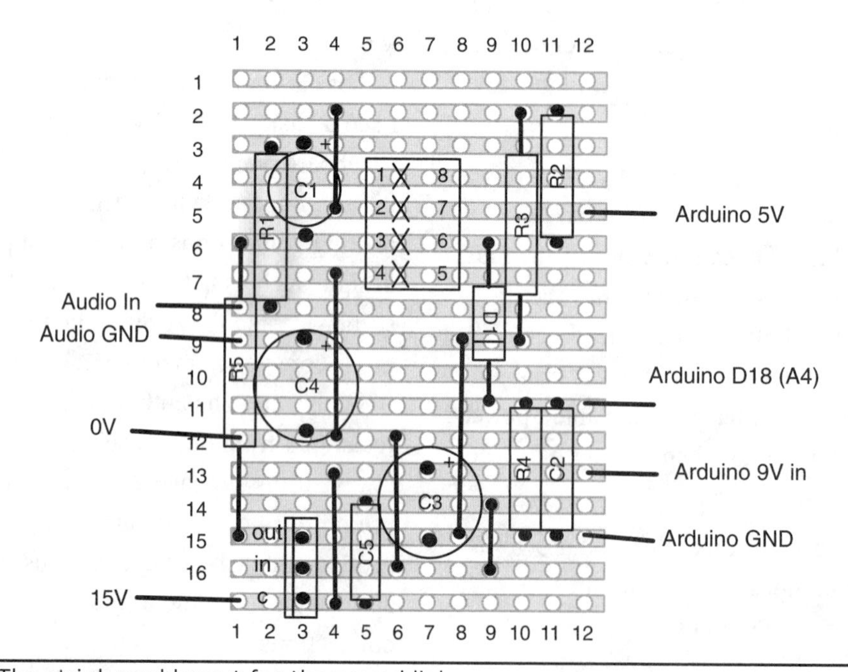

Figure 7-4 The stripboard layout for the sound link

Notice that some of the copper strips are cut where the IC is to be installed. These places are marked by an "X" on the stripboard layout of Figure 7-4. Such cuts in the strips are made by taking a drill bit, setting the tip into the hole where you want to make the cut, and twisting it between your fingers a few times to remove the copper track without actually making much of a hole.

So, start by cutting a piece of stripboard that is 17 strips long, each strip having 12 holes to it. You can do this with a strong pair of scissors, but this will usually result in some ragged edges. Greater neatness can be achieved by scoring the board with a craft knife and breaking it over the edge of a table. But be careful when the stripboard breaks, it can leave sharp edges.

Using Figures 7-4 and 7-5 as a guide, make the four cuts in the track with a drill bit.

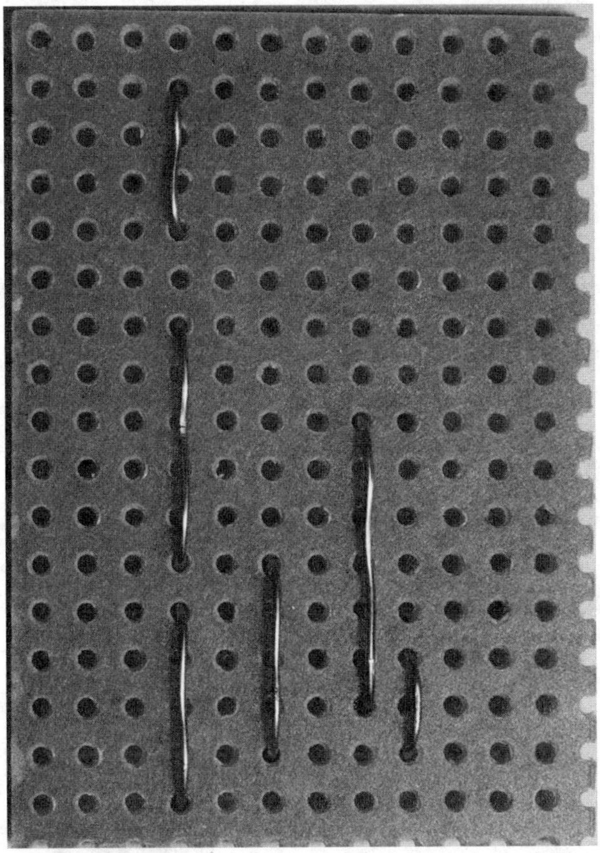

Figure 7-6 The stripboard with links

Step 2. Solder the Links

The rule with soldering to stripboards—or for that matter, PCBs—is to solder the lowest components first. This is so that if you put the part in place and turn the board onto its back, the weight of the board will keep the component in place while you solder it.

In this project, we have six links to make on the board, and we are going to make these links first.

Using Figures 7-4 and 7-6 for reference, cut solid core wire to roughly the right lengths, solder them in place, and cut off the excess wire.

Step 3. Solder the Resistors and Diode

The next lowest components are the resistors and diodes (Figure 7-7). Do not solder the R5

Figure 7-7 The stripboard with resistors and diode

resistor yet, this is added after we have attached the trailing leads.

The resistors can go either way around, but when you solder the diode into place, it must be as shown in Figure 7-7, with the striped end toward the bottom of the board.

Step 4. Solder the IC

You may choose to use an IC socket to remove the possibility of damaging the IC by overheating during soldering. If you decide to solder the IC directly, try to minimize the time that you are heating the pins. Also, if you have a difficult connection that takes a while for the solder to flow, wait until the IC is cool again before soldering the next lead.

Make sure the IC is the right way around. The little dot denotes pin 1 and should be at the top left. See Figure 7-8.

Step 5. Solder the Remaining Components

Once the IC is in place, only the four capacitors and the voltage regulator remain to be soldered. Electrolytic capacitors are polarized, which means we need to get them the right way around. The longer lead will be the positive lead. The negative lead will usually have a diamond symbol next to it.

The voltage regulator has the metal tab toward the left side of the board.

Figure 7-9 shows the completed stripboard.

Note that we have not attached the resistor R5 yet. This will be soldered into place after we have attached the leads to the board.

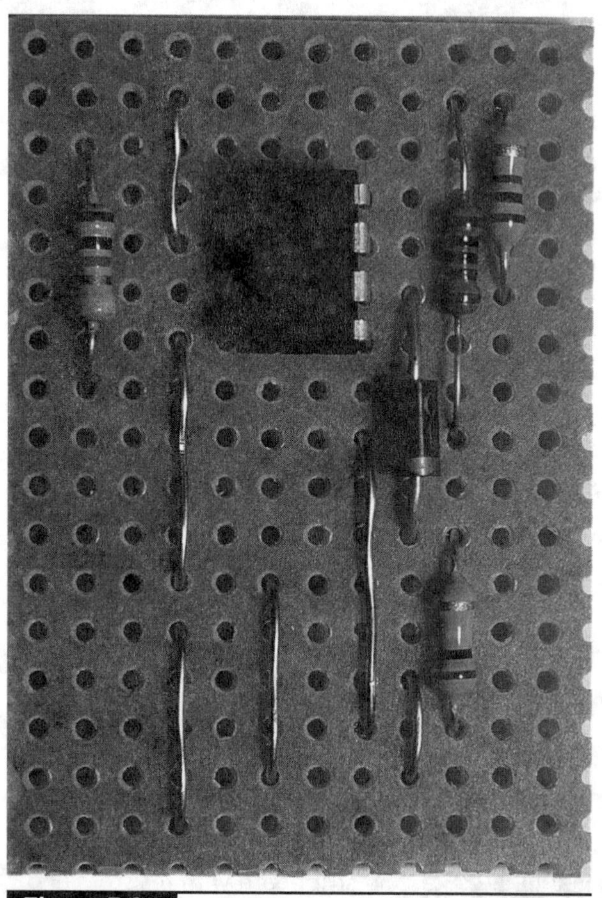

Figure 7-8 The stripboard with the IC in place

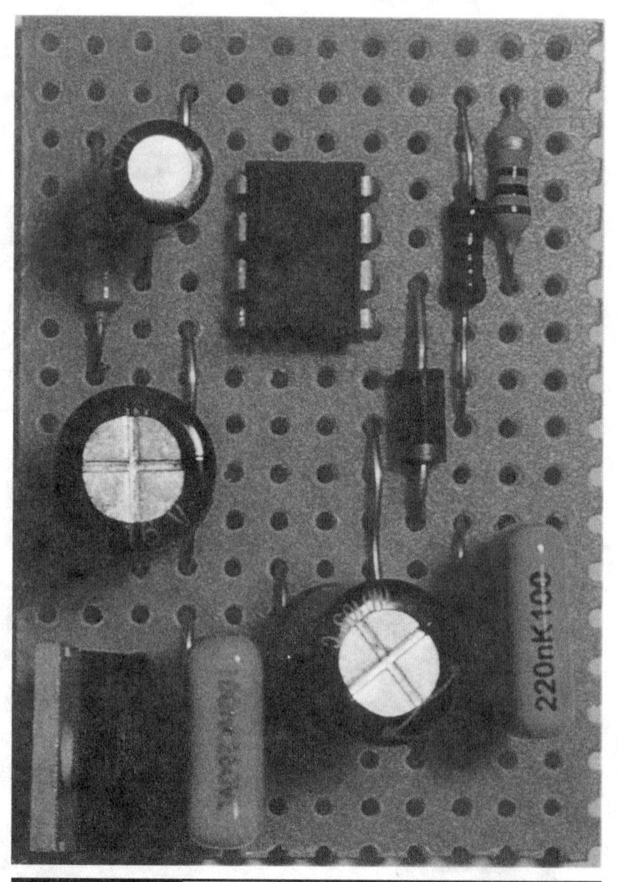

Figure 7-9 The completed stripboard

At this point, a thorough inspection of the board is a good idea. Inspect the top of the board to make sure all the components and links are in the places indicated by Figure 7-4, and look at the bottom of the board to make sure all the breaks are in the right places and that there are no accidental bridges of solder between tracks.

Step 6. Attach Trailing Leads

We need to be able to connect our sound interface to both the Android tablet and the Arduino board, so the next step is to solder some leads to the board (Figure 7-10).

Starting with the audio lead, you have two choices here. Either buy a plain stereo 3.5mm jack plug and solder some leads to it, or buy or scavenge a cheap or unwanted pair of headphones and chop the lead off, about 10 inches (250mm) from the plug. That way, you have a ready-made lead. The lead will probably have an outer shielding layer of braided wires and a pair of inner wires, often with red and white insulation for the left and right channels from the Android device's sound card. We only need one of the channels, so you can cut off one of the inner leads (let's say the white one). When you do this, make sure there is no bare wire sticking out that could short to the braided outer shielding since this could damage the Android device.

Gather the outer shielding wire and twist it together, and then apply some solder into it.

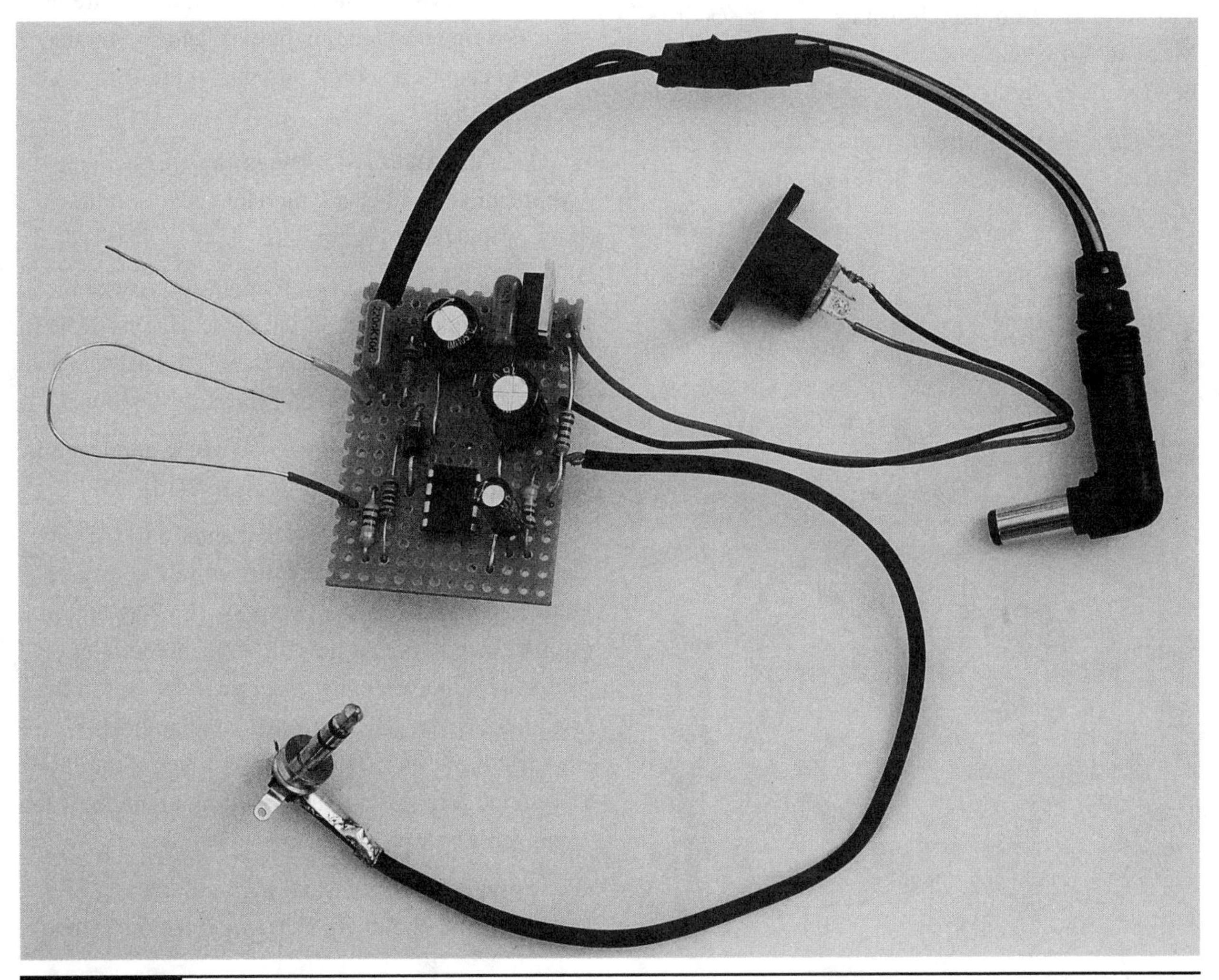

Figure 7-10 The stripboard with trailing leads attached

Carefully strip the remaining inner red lead and flow some solder into that, too.

Push the shielding wire and red wire through the holes in the stripboard and solder the underside, as shown in Figures 7-10 and 7-4.

In addition to the audio lead, we will need a lead connected to a socket to take power from the 15V power supply, and another lead connected to a low-voltage power plug to supply power to the Arduino.

The socket lead is made by soldering short lengths of multi-core wire between the sound interface card and a 2.1mm power socket. However, for the power plug, we can either make ourselves a lead using a plug and multi-core wire, or chop the end off an unwanted power supply lead.

We can now solder the R5 resistor into place above the leads on the left-hand side of the board. Make sure it is not touching the exposed leads, however. See Figure 7-11.

Finally, we need wires to connect the 5V supply from the Arduino back to the sound interface card

Figure 7-11 The board with the R5 in place

and from the sound card to connection A4 of the Arduino board.

Step 7. Test

That completes the construction of the board. Now we just need to test it.

Rather than test the full home automation software, we are going to download a test application for the Android device and a test sketch for the Arduino from the book's web site at www.duinodroid.com.

Let's start by installing the Android application onto our Android device.

Unlike the iPhone, you can download your Android applications from anywhere you like. This does mean that you have to make sure you are not downloading anything malicious, and so you may need to change a setting on your Android device to accomplish this.

Open the Android "Settings" app, navigate to Applications, and check the Unknown Sources box, as shown in Figure 7-12.

To actually install the test application, open the browser app on your Android device and navigate to www.duinodroid.com. Click the Downloads tab and then click the link for Sound Interface Test App.

This will initiate a download of the application, after which you can run it (Figure 7-13).

The test application lets you enter a four-digit hexadecimal (hex) number that will be converted into sound pulses when you press the Play button. It is these pulses that the full home automation software will eventually generate to be sent to the Arduino, which will reconstruct the number. We can test it out on its own without plugging it into the sound interface by just entering a number (each digit must be 0 to 9 or A to F).

Try entering "0000" and play, and then "FFFF" and play. You should be able to tell the difference

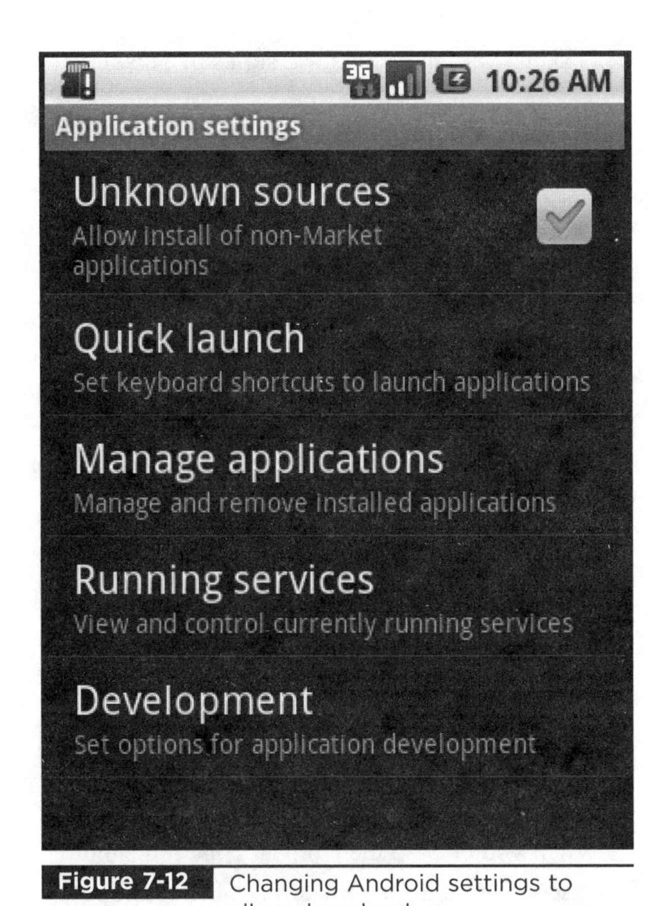

Figure 7-12 Changing Android settings to allow download

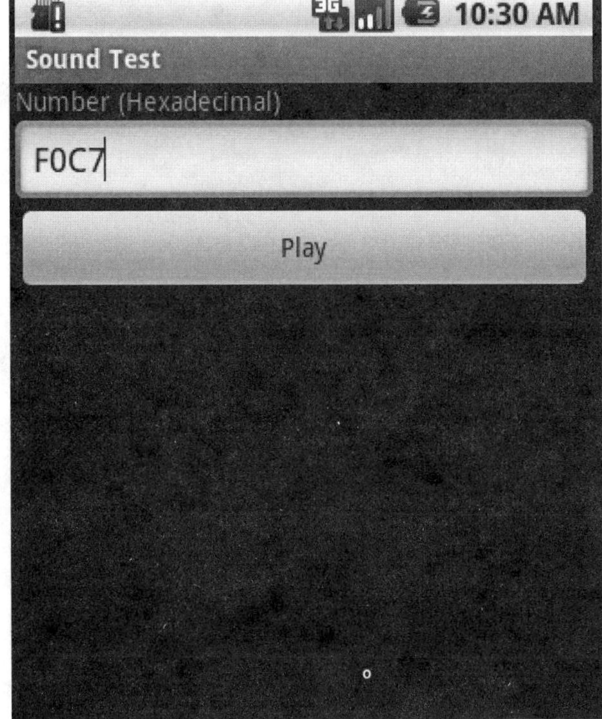

Figure 7-13 The Android test app for the sound interface

between the two sets of warbling coming through the Android device's speakers.

We now need to set up our Arduino environment so that we can complete the link and have our Arduino tell us the numbers we are sending it.

The Arduino board we are using (Arduino Uno) uses a special-purpose development environment that lets us send programs, or "sketches" as they are called in the Arduino world, to the board through the USB lead.

We need to install the Arduino environment, and rather than repeat instructions given elsewhere, please refer to the official Arduino site (www.arduino.cc) and follow the instructions there for installing the Arduino environment on your computer. You will find separate instructions there for Windows, Linux, and Mac. This book uses version 22 of the Arduino software and the Arduino Uno interface board; however, you should have no problem using later versions of Arduino.

Once your Arduino environment is set up, you need to install the test sketch for the project. In fact, all the sketches for the projects in this book are available in a single zip file that can be downloaded from www.duinodroid.com.

Unzip the file and move the entire Arduino Android folder to your sketches folder. In Windows, your sketches folder will be in My Documents/Arduino. On the Mac, you will find it in your home directory under Documents/Arduino/, and on Linux it will be in the sketchbook directory of your home directory.

Restart the Arduino software and then from the File menu, select Sketches, Arduino Android, and then sound_test. This will open the Sound Test sketch shown in Figure 7-14.

Figure 7-14 The Sound Test sketch

In the "Theory" section at the end of this chapter, we will look at how this sketch decodes the warbling coming from the Android device. For now, however, we will be content just to use it.

This sketch simply waits for a series of 16 sound pulses and then echoes the number it receives in the Serial Monitor of the Arduino software. In other words, we are going to use our computer as a monitor to tell us what the Arduino is receiving from the Android device.

Connect your Arduino board to your computer via USB and upload the sketch to the board by clicking the upload icon (second to right on the toolbar). If you get an error message, you will need to set the type of board you are using and the connection. To set the board, select the Tools menu and then the Board option. This will give you a list something like that in Figure 7-15.

Select the option for the type of board you are using. Then, do the same thing for the serial port, which is also on the Tools menu. This will generally be the top option on the list of ports, and normally COM4 on Windows.

When you clicked the Upload icon, you should have seen some furious flickering of the little red

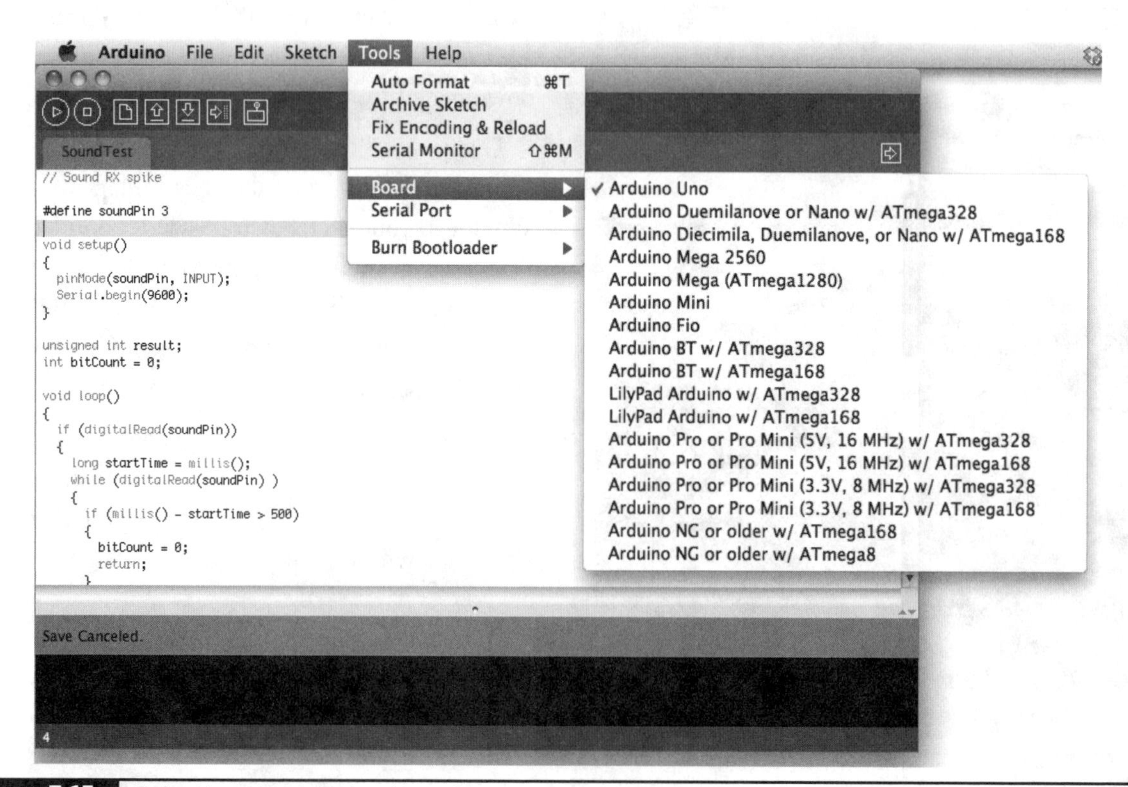

Figure 7-15 Selecting the Arduino board type

built-in LEDs on the Arduino board. This indicates that the Arduino board is receiving the sketch.

So, let's connect up our sound interface board and give it a go. Figure 7-16 shows the sound interface connected to the Arduino board. In particular, to the A4 and +5V connections.

Plug the audio jack into the sound card, and the 15V power supply into the socket on the training leads from the sound interface card. Then, plug the 9V plug from the sound interface board into the Arduino and attach the 5V and digital signal wires from the Arduino, as shown in Figure 7-16.

Now that everything is connected, we will enter a number on the Android test application and send it to the Arduino through the sound interface. The Arduino will then forward the number on to your computer over the USB lead attached to the Arduino.

To see the messages coming from the Arduino, open the Arduino Serial Monitor on your computer (Figure 7-17). This lets you see any serial communications coming from the Arduino.

Enter a number—say, F0C7—into the Android test application and press Play. If you can hear the warbling, you do not have the audio lead plugged into the Android tablet, so plug it in and try again.

Almost immediately, you should see the following message appear in the Serial Monitor:

Figure 7-16 Wiring up to test the sound interface

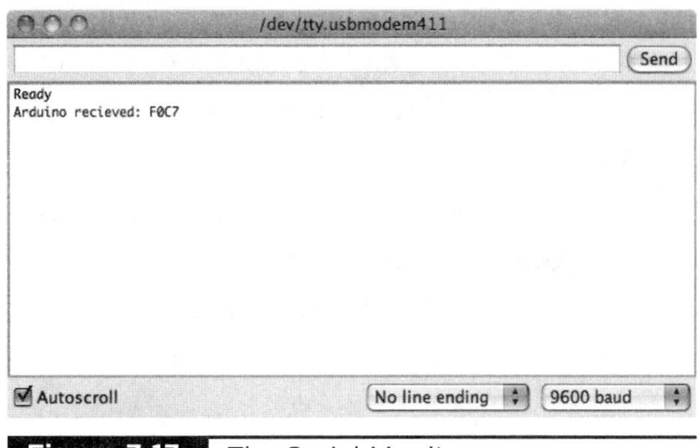

Figure 7-17 The Serial Monitor

"Arduino Received: F0C7." If you got a different number, or no number at all, then try adjusting the volume level on the Android tablet. You will probably need to set the volume to about three quarters of maximum.

Step 8. Box the Controller

One of the goals that the author had in mounting the Android tablet into the box was to avoid damaging it, so that if at some point it needed to be extracted from the home automation controller, this would be possible.

The most difficult part of making a box for the controller is cutting a big window in the box lid to allow access to the Android tablet's touchscreen. This is made a lot easier if you have a rotary hand tool like the Dremel, which has a cutting disk.

As the saying goes "measure twice, cut once," so carefully work out where you need to cut your window in the lid and mark it with pencil. Once you are absolutely sure everything is in the right position, cut out the window (Figure 7-18).

Note that the author has also cut out a V-shaped corner to allow the Back button to be pressed on the tablet.

We need a few other holes in the box itself, including one for the Android tablet's power lead. This is supplied through a hole on the side of the box that is big enough for the plug to pass

Figure 7-18 The box lid

completely through without the need for modification.

We also need a small hole next to the on switch for the Android tablet. This only needs to be big enough to poke a screwdriver through to turn the tablet on.

The remaining hole that we need in the box is for the power socket that we attached by flying leads to the sound interface board.

The exact position of these holes will depend on the size of your box and of the Android tablet itself. In the next chapter, we will also make some more holes in the board to fix the perfboard containing the power control electronics, and optionally a hole to allow a USB connector to be attached to the Arduino board. So you may wish to wait until the next chapter before finalizing your box construction. Unfortunately, most Evil Geniuses are not known for their patience, so you too may "need" to see how it will all look in a box.

To fix the tablet to the lid, a length of drilled constructional metal from a project kit was used. This material is available from hobby shops and is useful because it can be easily bent and cut into the shape you need. The material was bent to fit over the back of the tablet, and the last holes drilled turned out to be big enough for the center screws that held the lid on to pass through and into the base of the box. See Figure 7-19.

Figure 7-19 The tablet fixings

We will add more to this project box as we proceed through the home automation section of this book.

Android Software

Now that we have proved the sound link and fitted the tablet into a box, it is time to look at the Home Automation software. Like the "sound test" app, this can be downloaded from www.duinodroid.com, where the source code for both applications is also freely available if you want to modify the software for your own ends. Once downloaded and installed, leave the sound interface and Arduino attached so you can see the messages the Android tablet is sending.

The structure of the Android software (Figure 7-20) is unusual in that it actually contains a small web server. This web server serves HTML to the device's browser, so that the main user interface to the application is a web browser running on the same Android tablet as this mini web server.

This means that other devices on the local network can also connect to the Android tablet, and if you configure your home wireless modem correctly, you can also communicate with the Home Automation app over the Internet.

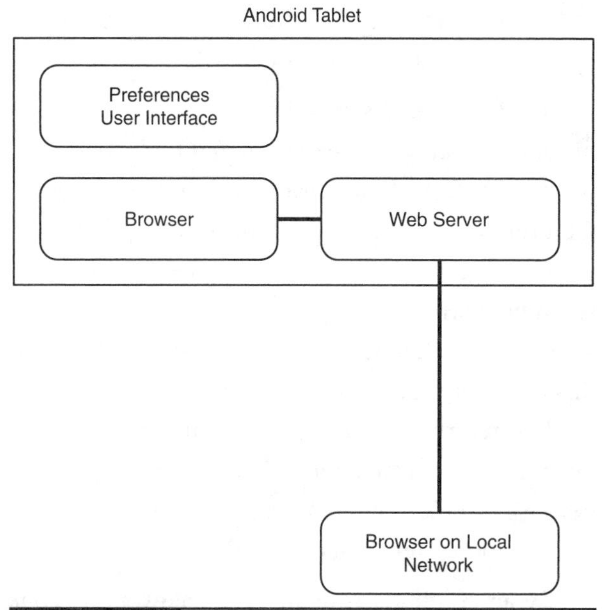

Figure 7-20 The structure of the Android app

A word of caution is necessary here. If you do allow Internet access, there is almost nothing in the way of security to stop someone who stumbles on your home automation server to take control of your house. So, you may prefer not to take this risk.

Figure 7-21 shows the first screen you see when the app starts.

The first line tells us that the server is running at the URL displayed. In this case, "http://10.0.2.15:8080/home?".

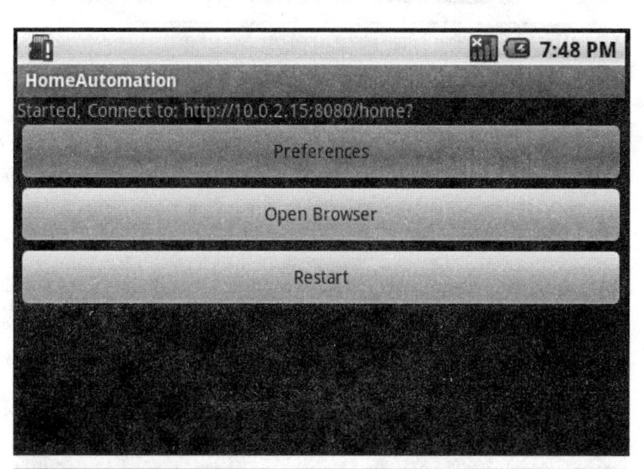

Figure 7-21 The Home Automation app start screen

But first, we will click Preferences, which will open a second screen (Figure 7-22).

The Preferences screen lets us assign meaningful names to the power and light sockets we will control. If you scroll down to the end of the Preferences screen, you will see an option to set a password. This password will be required when unlocking the door. Do not worry about setting the preferences now; you can come back to this any time and change them. For now, just click Back to return to the main menu, and then click Open Browser, which will launch the Android device's browser on the main page of the automation system (Figure 7-23).

If your computer is on the same network as your Android device, you can type the URL at the top of Figure 7-23 into the browser on your computer and see the same screen. This means you can control your home from any device connected to your home network using just a browser. That includes laptops, netbooks, iPhones, iPads, and iPod touches, as well as most other smartphones.

How cool is that?

From the main menu, we then have a series of different options for controlling our home. Unfortunately, none of these options will do anything yet, because we have not made the hardware to control the power, heating, and the door. But we can at least have a look at the options that will gradually become available as we work our way through this section of the book.

Clicking Power will open up our power and light control options (Figure 7-24).

From this screen, you can turn outlets and lights on and off individually, or turn all of them on and off. Remember, you can at any time go back to the Home Automation app itself and change the preferences to use different names for the channels. Simply click the Android Home button and bring the Home Automation app back to the foreground.

Try clicking On next for the first outlet. You should see a number appear in the Serial Monitor. Unplug the audio lead from the Android tablet, and you should hear the different warbles that you get when each button is pressed.

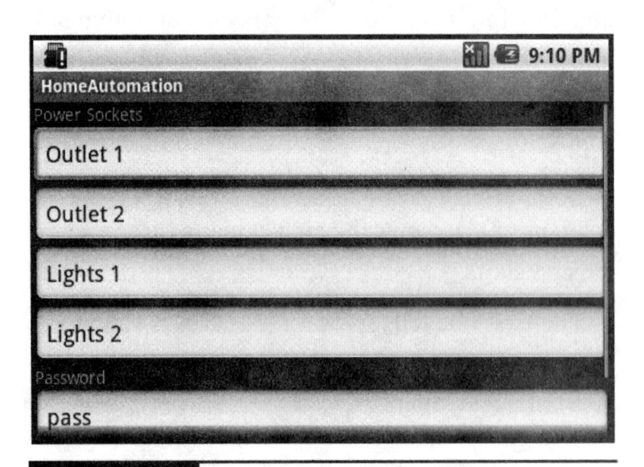

Figure 7-22 The Preferences screen

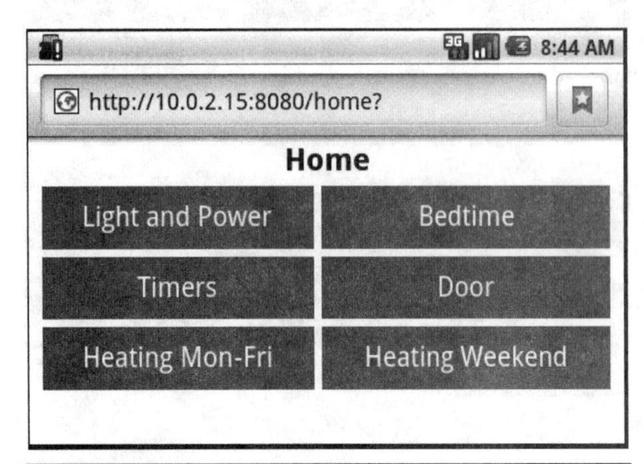

Figure 7-23 The main menu

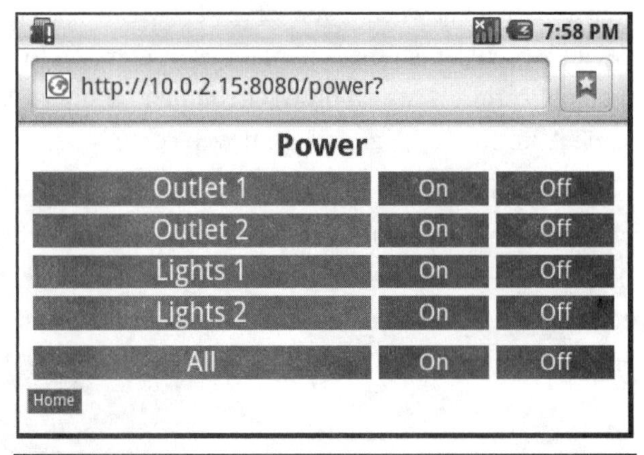

Figure 7-24 The power and lights

At the bottom of the browser window, you will see a green button labeled Home. Click this to return to the main menu and then click Bedtime (Figure 7-25). This interesting feature is designed to turn everything off when the Evil Genius is ready to go to bed. It allows 10 minutes from pressing the button until all the power and lights are turned off. This gives the Evil Genius enough time to mount his Segway and scoot the half mile to his bedroom before everything goes dark.

If the Evil Genius changes his mind, he can always click Back before the timer has expired and the lights and power will then not be turned off.

The next option from the main menu is Timers (Figure 7-26).

This option allows up to five timed events to take place. For each event, you can select a light or a power socket and turn it on at one time and then off at a second time. If you set the "off time" to a time that is before the "on time," then it assumes you mean the "off time to be for the next day.

The changes to the timers will not be saved until you click Save Changes.

Returning to the main menu, the next option is Door Lock (Figure 7-27).

In a vague nod in the direction of security, you must enter a password that matches the password set in Preferences before you can lock or unlock the door.

The final option in the main menu lets you set a temperature profile for the heating (Figure 7-28). The temperatures that you set for various times of day will be transferred to the Arduino and then onto the project of Chapter 9, a "smart" thermostat, using an Arduino-to–Arduino RF link. See Chapter 9 for more details of this feature.

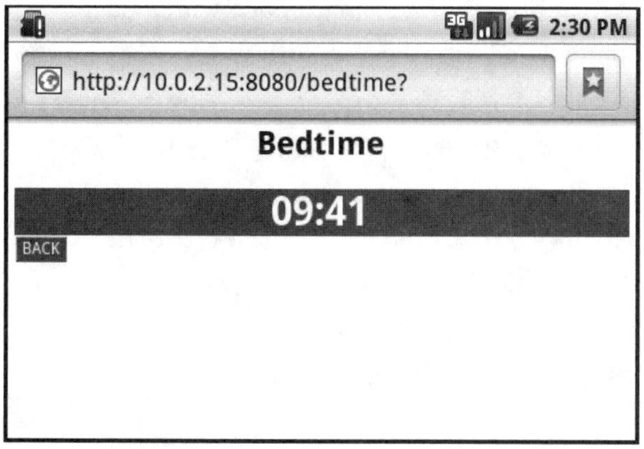

Figure 7-25 Bedtime

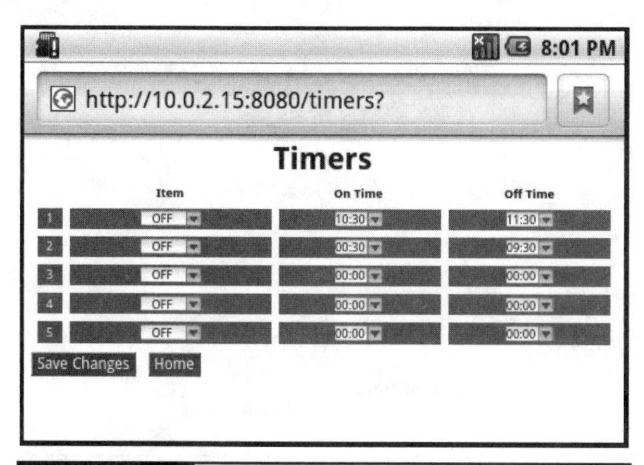

Figure 7-26 Timers

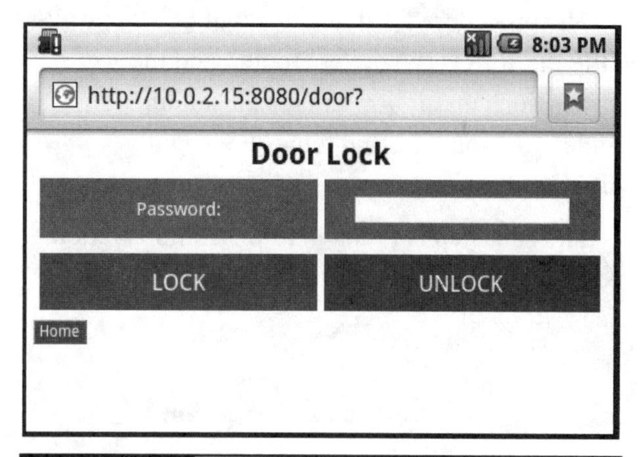

Figure 7-27 The Door Lock screen

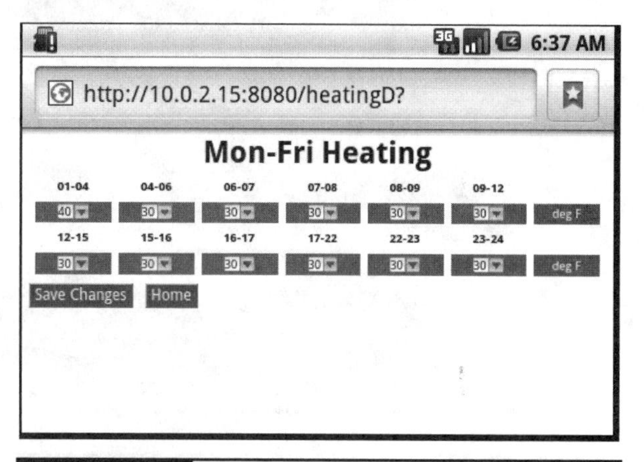

Figure 7-28 Heating

Internet Access

You have probably already tried typing in the URL for the home automation controller on a computer connected to your home wireless network. Remember that your Android tablet is running as a web server, so you can actually take this a step further and access your home automation controller from anywhere on the Internet. However, to do this, you will have to change some settings on your home wireless hub. This is because, by default, your wireless router does not allow any computers on your home network to be directly accessible from the big bad world of the Internet. To make this possible, you have to open up access to the Android tablet via port 8080.

Technically, what you are trying to accomplish is opening port 8080 to incoming HTTP requests to the Android Tablet.

The procedure for doing this will be slightly different for every wireless router, and the instructions that follow are for the author's router. So, you will find things are slightly different for you.

The first step is to log in to the administration console for your router from your browser.

From the main status page after logging in (Figure 7-29), you will see some numbers. Look for something like "external IP address" and make a note of the IP address. You will need this when connecting from outside your network.

You are now looking for an option resembling "NAT Virtual Server" (Figure 7-30). This will allow you to route any web requests coming to your router to the Android tablet. Notice the changes circled. The key things are to get the local IP address of the Android tablet correct. Note that this may change unless you have set the DHCP lease time to be as many days as possible. You will find this option on the DHCP settings page of your home hub.

Once you think you have made the necessary changes to your router, you will only be able to test it by asking someone outside your home network to try it. For the ultimate Evil Genius wow-factor, use the browser on a smartphone with a data contract and turn the WiFi off.

So if your external IP address is 98.76.543.21, enter the URL http:// 98.76.543.21:8080/home? into the address bar of your phone's browser. You should see the home page appear on your screen. If you think that's exciting, just wait until you can actually turn things on and off!

Service Information

LAN Interface

IP Address	Subnet	MAC Address	Status
192.168.1.1	255.255.255.0	00:21:63:3b:c5:00	✓
Port	**Speed**	**Duplex**	**Status**
1	-	-	✗
2	-	-	✗
3	-	-	✗
4	-	-	✗

WAN Interface

PVC	VPI/VCI	IP Address	Subnet	Gateway	Encapsulation	Status
PVC-0	0 /38	[illegible]	255.255.255.255	92.16.0.1	Route -PPPoA	✓

Figure 7-29 The router status page

NAT - Virtual Server

NAT - Virtual Server	
Virtual Server for	Single IP Account
Rule Index	1
Application	Home Automation
Protocol	TCP
Start Port Number	8080
End Port Number	8080
Local IP Address	192.168.1.4
Start Port(Local)	8080
End Port(Local)	8080

Figure 7-30 The NAT Virtual Server setup

Theory

This is a complex project and there are lots of interesting theoretical concepts to explore.

Encoding Data as Sound

When the Evil Genius was nothing more than a Slightly Naughty Genius, he had a home computer. This was a source of constant amusement for him. The computer did not have a hard disk, nor even a floppy disk, it just had some wires that led to a battered old cassette tape recorder.

To load a game into the computer, the Evil Genius had to first put a cassette tape into the tape player, rewind it and press Play (while the Pause button was depressed). He then had to type LOAD into his computer, release the Pause button, and then press the ENTER key once the tape leader had grunched its way past the tape head. There would then be a series of strange warblings, possibly for several minutes, before the massive 3-kilobyte program had finished loading.

This strange warbling is the data for the program encoded as a series of tones. We use this exact same approach when getting our Android tablet to communicate with the Arduino.

Figure 7-31 shows the oscilloscope trace of just 16 bits of data encoded as sound.

This trace is the raw output from audio output. You can make out 16 pulses: four long, four short, two long, three short, and then three long. The short pulses represent a 0 and the long pulses a 1. So, if we write out the number in binary, we get 1111 0000 1100 0111, which in hexadecimal is F0C7, the number that we sent.

If we zoom in on the signal (Figure 7-32), we can see that each pulse is actually made up of a sine wave at a higher frequency, modulated into pulses. Looking closely at the signal, we can see that the period for one cycle is exactly 1 division or 1ms, so the frequency of this signal is 1 kHz. Looking back at Figure 7-30, you can see that we

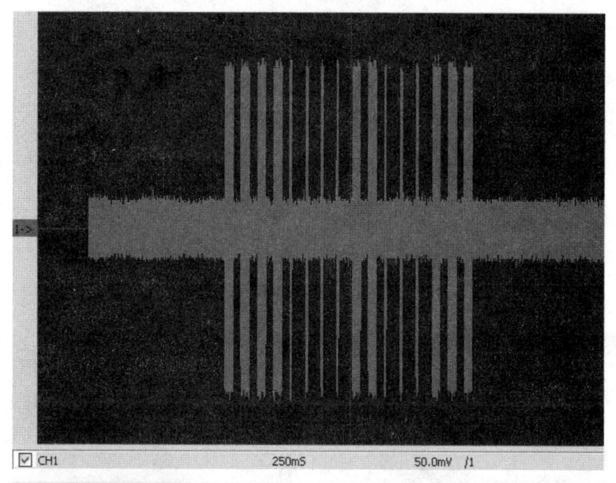

Figure 7-31 An oscilloscope trace of output from the Android tablet

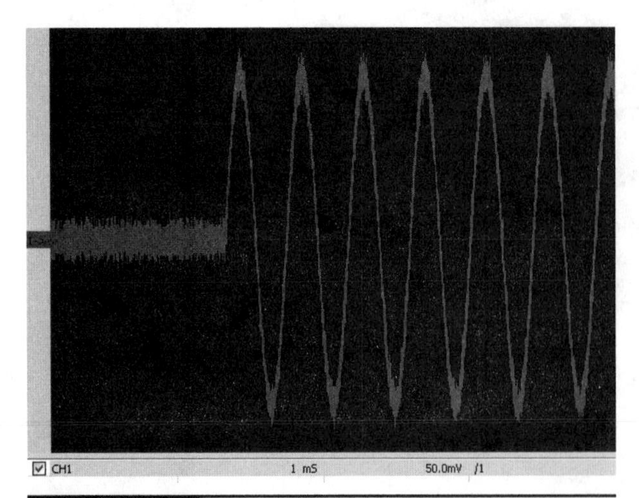

Figure 7-32 Zooming in on the signal

can fit about 4 pulses into a 250 ms division, so each pulse has to fit into a gap of about 62 ms.

When we look at the Android code for generating the signal, we will see that this figure is actually 64 ms, and that a long pulse is 32 ms and a short pulse 8 ms.

As well as downloading the Android app that's ready to use, the source code for both the Sound Test app and the full automation controller are available as open-source software under a GPL license. You can download the source-code files from www.duinodroid.com.

The core class that generates the correct sound pulses is called Beeper, and the listing for this is given next:

```
import android.media.AudioFormat;
import android.media.AudioManager;
import android.media.AudioTrack;

public class Beeper {

  private final static int SAMPLE_RATE = 16000;

  private final static int ONE_DURATION = 32;
  private final static int ZERO_DURATION = 8;
  private final static int BIT_DURATION = 64;
  private final static int DURATION = BIT_DURATION * 32;
  private final static float f = 1000.0f;

  private short[] buffer = null;

  public void beep(int word) {
    AudioTrack at;
    int bufsizbytes = DURATION * SAMPLE_RATE / 1000;
    int bufsizsamps = bufsizbytes / 2;
    buffer = new short[bufsizsamps];
    fillbuf(word, bufsizsamps);
    try {
      at = new AudioTrack(AudioManager.STREAM_MUSIC, SAMPLE_RATE,
          AudioFormat.CHANNEL_CONFIGURATION_MONO,
          AudioFormat.ENCODING_PCM_16BIT, bufsizbytes,
          AudioTrack.MODE_STATIC);
      at.setStereoVolume(1.0f, 1.0f);
      at.write(buffer, 0, bufsizsamps);
```

```
        at.play();
      } catch (IllegalArgumentException e) {
        e.printStackTrace();
      }

    }

    void fillbuf(int word, int bufsizsamps) {
      double omega, t;
      double dt = 1.0 / SAMPLE_RATE;
      t = 0.0;
      omega = (float) (2.0 * Math.PI * f);
      for (int i = 0; i < bufsizsamps; i++) {
        if (toneRequired(t, word)) {
          buffer[i] = (short) (32000.0 * Math.sin(omega * t));
        } else {
          buffer[i] = 0;
        }
        t += dt;
      }
    }

    boolean toneRequired(double t, long word) {
      int ms = (int) (t * 1000);
      int bitIndex = ms / BIT_DURATION;
      int bit = (int) ((word >> (15 - bitIndex)) & 1);
      int msWithinBit = ms - (bitIndex * BIT_DURATION);

      if (bit == 1 && msWithinBit < ONE_DURATION) {
        return true;
      }
      if (bit == 0 && msWithinBit < ZERO_DURATION) {
        return true;
      }
      return false;
    }

  }
```

The constants at the top of the file define the frequency and pulse durations that the system is going to use. This class sound interface system could easily be extended to allow much faster data rates and greater quantities of data, by increasing the carrier frequency (f), the sample rate (SAMPLE_RATE), and decreasing the pulse sizes. However, we do not have much to send, so I have chosen reliability over performance. Note that if you do change these parameters, you will probably also need to change the electronics, too.

The only public method on the class is called "beep" and this takes the 16-bit number to be sent and converts it into the series of pulses.

To generate the pulses, we actually have to write the samples that make up the waveform into a buffer, and then play the contents of the buffer as if it were a sound file. The method "fillBuffer" does most of this work, generating the sine wave using the trig sin function, but only when the "toneRequired" method says that there should be a tone. It does this by working out which bit we are supposed to be beeping, and then returning true if the time into that bit is less than the duration for the bit, depending on whether it is a 1 or a 0.

The Sound Interface Electronics

The schematic diagram for the audio interface is repeated in Figure 7-33, with a test point "A" marked on it.

The job of the electronics is to take the small signal from the Android device—probably about 300mV peek to peek (Figure 7-30)—and produce a series of logic level (0–5V) pulses for the Arduino. It accomplishes this in a series of stages as summarized in Figure 7-34.

The first step is to amplify the signal from the Android tablet. We do this using an op amp IC. This little amplifier on a chip is arranged to have a gain of about 1000. This means that the signal at the output will be 1000 times the signal at the input. Well, if that were actually the case, the output would be 300V, but what happens is that the output is limited to the voltage supplying the op

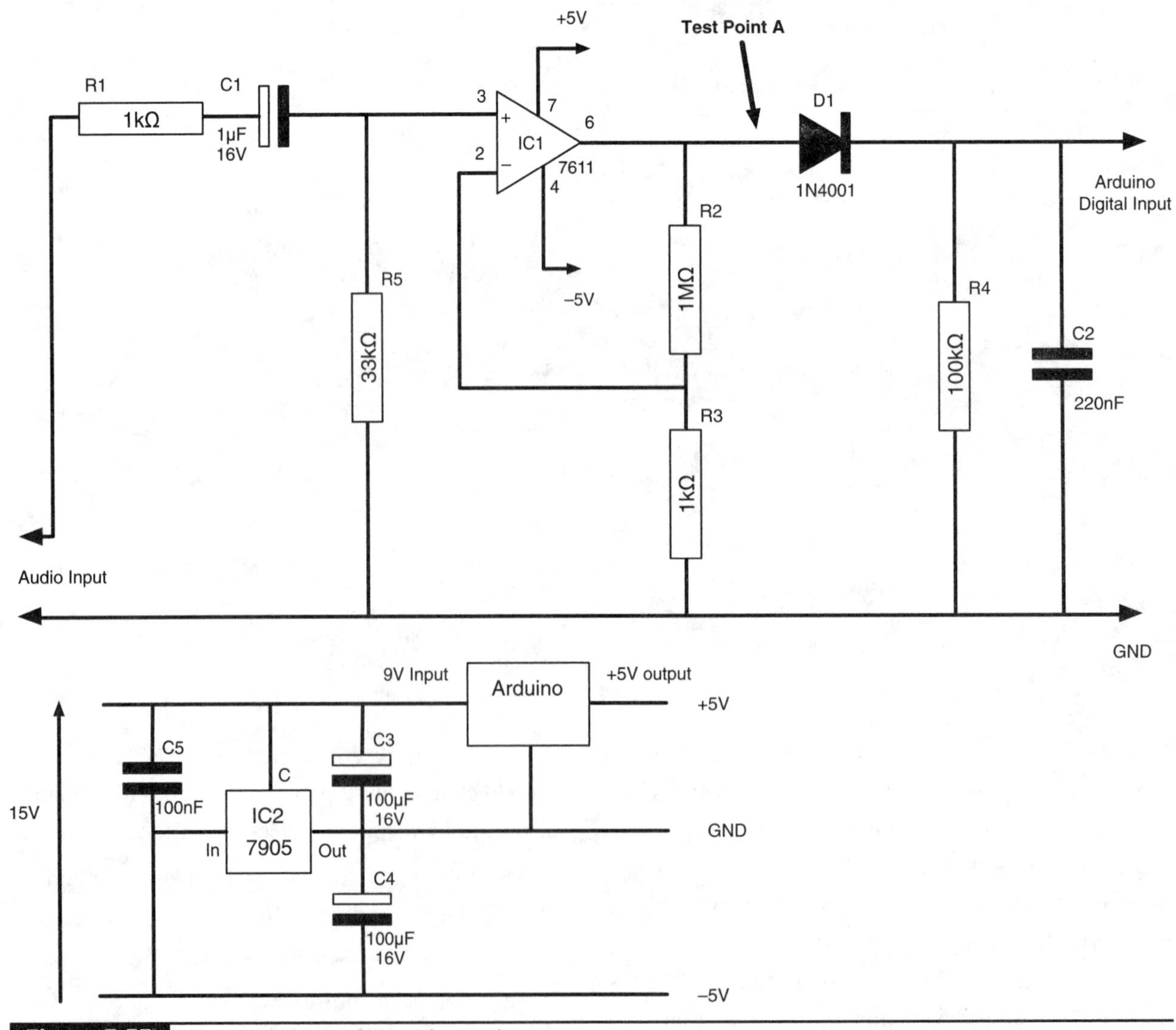

Figure 7-33 The sound interface schematic

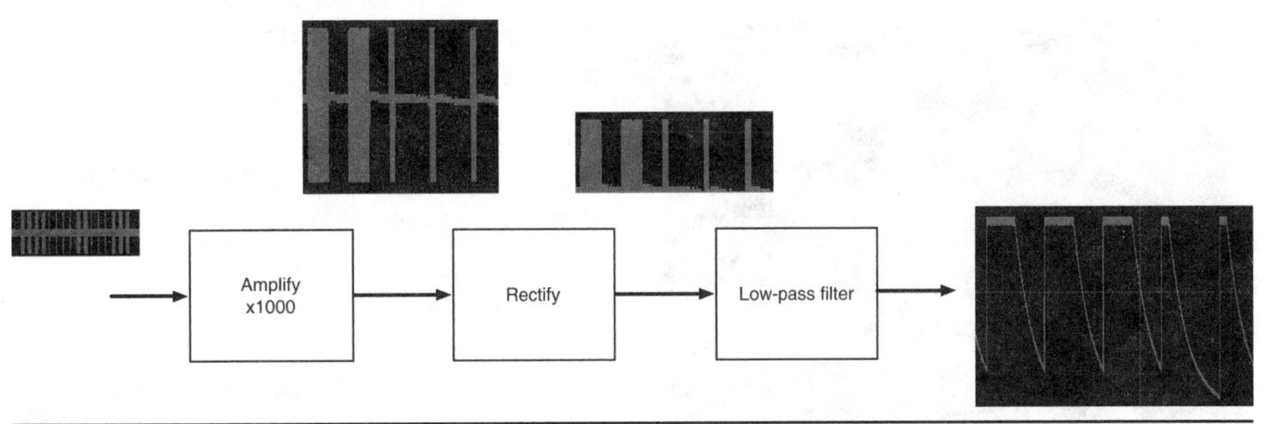

Figure 7-34 A block diagram for the electronics

amp, which in this case is +–5V. So, in effect the signal will be amplified as much as possible. So that if the signal from the Android tablet is quite weak, it should still be amplified to a decent level.

Before we move onto the next stage, it is worth looking at how the circuit is supplied with power. The signal from the Android tablet swings either side of the ground connection, and to amplify it, the op amp needs what is called a "split" power supply. That is, it needs ground, +5V and –5V. By a cunning use of the Arduino's built-in 5V voltage supply, we manage to contrive this arrangement of +–5V and a 9V supply for the Arduino itself.

Referring to the bottom half of Figure 7-33, we can see that our basic input supply comes as 15V from a wall-wart power supply. The most negative side of this supplies the –5V for the op amp. We then use a negative 9V regulator to provide a regulated voltage 9V less than the 15V, which will be our GND for the op amp and Arduino. This 9V drop also supplies the Arduino, and the regulated +5V from the Arduino's own voltage regulator provides the +5V for the op amp.

The next stage after amplifying the signal is to rectify it. Our Arduino could be damaged if we try and apply negative voltages to an input pin, so we need to make sure that the voltage is always somewhere between GND and +5V. The diode accomplishes this task. However, we still have the problem that each pulse is actually made up of a 1-kHz wave, and we need to remove that signal so we are just left with the "envelope" of the pulses. We use a low-pass filter to accomplish this. This just comprises the resistor R4 and C2. This also has a bit of a distorting effect on the signal, but it doesn't matter.

Figure 7-35 shows the oscilloscope trace of the signal. The top trace shows the amplified signal at test point A and the bottom trace the signal after rectification and low-pass filtering.

Figure 7-36 is a close-up of the final signal. The Arduino digital input will treat everything above the 3V line as a 1, and everything below as a 0. So at that level, a long pulse will have a duration of about 40 ms, and a short pulse about 20 ms.

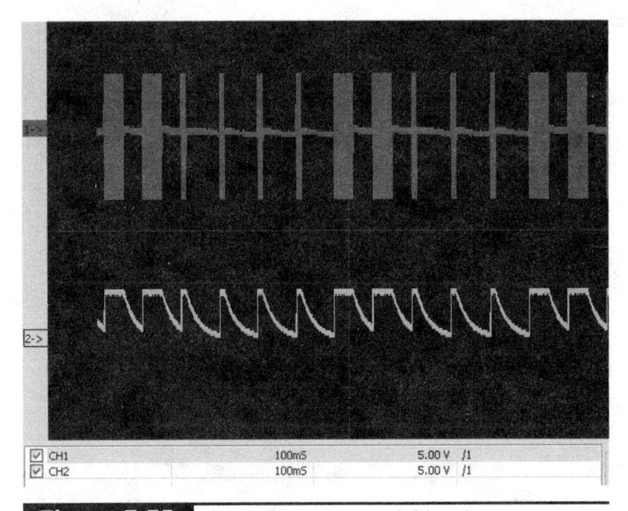

Figure 7-35 Rectification and filtering

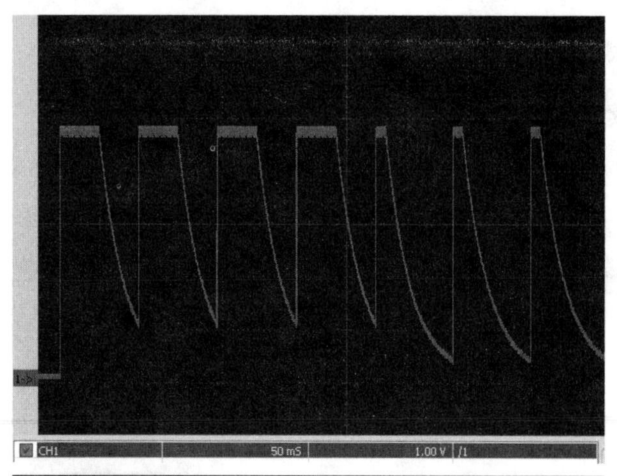

Figure 7-36 Close-up of the final signal

Decoding the Sounds on the Arduino

Our signal is at last in the Arduino, and we just need a way of decoding the pulses of varying length into a number.

The test sketch for this is shown in the following listing.

```
#define soundPin 18
#define zeroDurationFrom 10000
#define zeroDurationTo 25000
#define oneDurationFrom 35000
#define oneDurationTo 50000
#define resetTimeout 3000

void setup()
{
  pinMode(soundPin, INPUT);
  Serial.begin(9600);
  Serial.println("Ready");
}

unsigned int result;
int bitNo = 0;
long lastPulseTime = 0;

void loop()
{
  long pulseLength = pulseIn(soundPin, HIGH, oneDurationTo * 2);
  long timeSinceLastPulse = millis() - lastPulseTime;
  lastPulseTime = millis();
  if (pulseLength == 0 || timeSinceLastPulse > resetTimeout)
  {
    bitNo = 0; result = 0;
  }
  else
  {
    if (pulseLength >= zeroDurationFrom && pulseLength <= zeroDurationTo)
    {
      result = result << 1;
      bitNo ++;
```

```
      }
      else if (pulseLength >= oneDurationFrom && pulseLength <= 50000)
      {
        result = (result << 1) + 1;
        bitNo ++;
      }
      else
      {
         Serial.print("Error pulseLength="); Serial.println(pulseLength);
      }
    }
    if (bitNo == 16)
    {
      Serial.print("Arduino received: ");
      Serial.println(result, 16);
      bitNo = 0; result = 0;
    }
  }
```

The first thing that we do in the sketch is define a few constants for the maximum and minimum pulse lengths. There is also a timeout set after which the software gives up on reading the number and starts again from the first bit.

Any Arduino "sketch," as programs are called in the Arduino world, will contain a setup and a loop function. The setup function is run once when the Arduino resets, and "loop" will be repeatedly indefinitely.

The setup function defines the pin (A4) to be the pin on which we are going to receive the pulses. It then initializes the serial connection so that the Arduino can send messages over USB to the Serial Monitor, and finally sends the message "Ready" to the computer.

Next, we define a few variables. The variable "result" will eventually contain the number sent to the Arduino. The value of "result" is built up literally bit by bit, and the variable bitNo is used to keep count of how many bits have been read. Finally, lastPulseTime contains a timestamp for when we last received a pulse. This allows the mechanism to timeout if no pulse has been received for a while.

Most of the action in this sketch takes place in the loop function. This function will be called repeatedly. Each time it is called, the first thing that happens is that the pulseIn function in the Arduino library is called. This waits for a pulse, but if none is received within the time period specified in the third argument (in this case, twice the duration of a "one"), the function returns 0.

The next lines test to see if a timeout has occurred. This could be because pulseIn has timed out, or because the time since the last pulse has been exceeded. If this is the case, then we start again from the top and the bitNo and result variables are reset.

Otherwise, we test the duration of the pulse to decide if it's a binary 1 or a binary 0 and modify the result variable and increment the bit count.

The last thing we do in the "loop" function is see if we have read all 16 bits, and if we have, it sends the resulting number to the serial port, where it will be passed to the Serial Monitor.

Summary

This concludes the project work for this first chapter. We now have a basic home automation controller. In the next chapter, we will make it actually do something useful: turn the power on and off on our appliances.

CHAPTER 8

Power Control

The Evil Genius used to use minion-based remote control. This worked by standing a minion next to each light switch, and when the Evil Genius required the light to be turned off, he would shout an order at the minion, who would turn the light off. A number of aspects of this technology were a little unsatisfactory, however—for example, when the light went out, the minions would sometimes fall asleep. Also, the minions were liable to wander off at an inopportune moment, leaving the Evil Genius barking orders that went unanswered, with dire consequences afterward for both the hapless minion and the Evil Genius' vocal chords.

The Evil Genius therefore decided that the first use of his Lair Automation system should be to control lights and power.

It is perfectly possible to buy remote control home automation units that connect to lighting and power sockets using the X10 protocol. These are, however, quite expensive, and being on a budget, the Evil Genius decided to use cheap, readily available RF (radio frequency) remote power sockets from his local supermarket (Figure 8-1). For around USD 25, this provides three or four remote control AC outlets and a remote control handset.

Note that the model shown is for the UK market, with the UK's unique style of mains outlet. Similar units are available for U.S., Asian, and continental European styles of outlet.

Figure 8-1 A remote control wall socket

We are going to modify the remote control handset so it can be driven by the Arduino of our home automation controller. In effect, the Arduino is going to be pressing the buttons.

We will use these cheap remote control outlets for both power outlets and lighting. For the lighting, we will need to modify the connections to them, to make them suitable for controlling domestic lighting.

Power Control Electronics

The Evil Genius has used a certain amount of cunning in producing this design. He is using the existing remote control for a power outlet and employing small reed relays to do the equivalent of pressing the buttons on the remotes.

This is all built onto a handy board that will allow our Arduino to control the remote for the power outlets. We could, if we wanted, use this module entirely independently of the Android tablet and just control the power from our computer using the USB connection. In fact, for test purposes, this is exactly what we will do, before we start connecting things up to the sound interface and Android tablet.

Figure 8-2 shows the assembled board. Note that it is on a large piece of "perfboard," because we are leaving room to add other parts as we proceed with the other home automation projects. "Perfboard" or "perforated board" is a prototyping board rather like the stripboard that we used in earlier projects, except that it is just board with holes in it at 1/10 inch pitch, without any copper strips on the back.

Constructing the Power Control Module

Figure 8-3 shows the schematic diagram for the power control module.

The Arduino has eight low-power reed relays attached to digital outputs. When one of the outputs is high, it energizes the relay coil, which closes the relay's switch. This switch is soldered across the terminals that would normally be closed when a button was pressed.

The use of relays might seem a little old fashioned, but a more sophisticated approach using transistors would depend too much on the type of remote used.

What You Will Need

You will need the components listed in the Parts Bin to make the power control module.

Figure 8-2 The power outlet remote control by Arduino

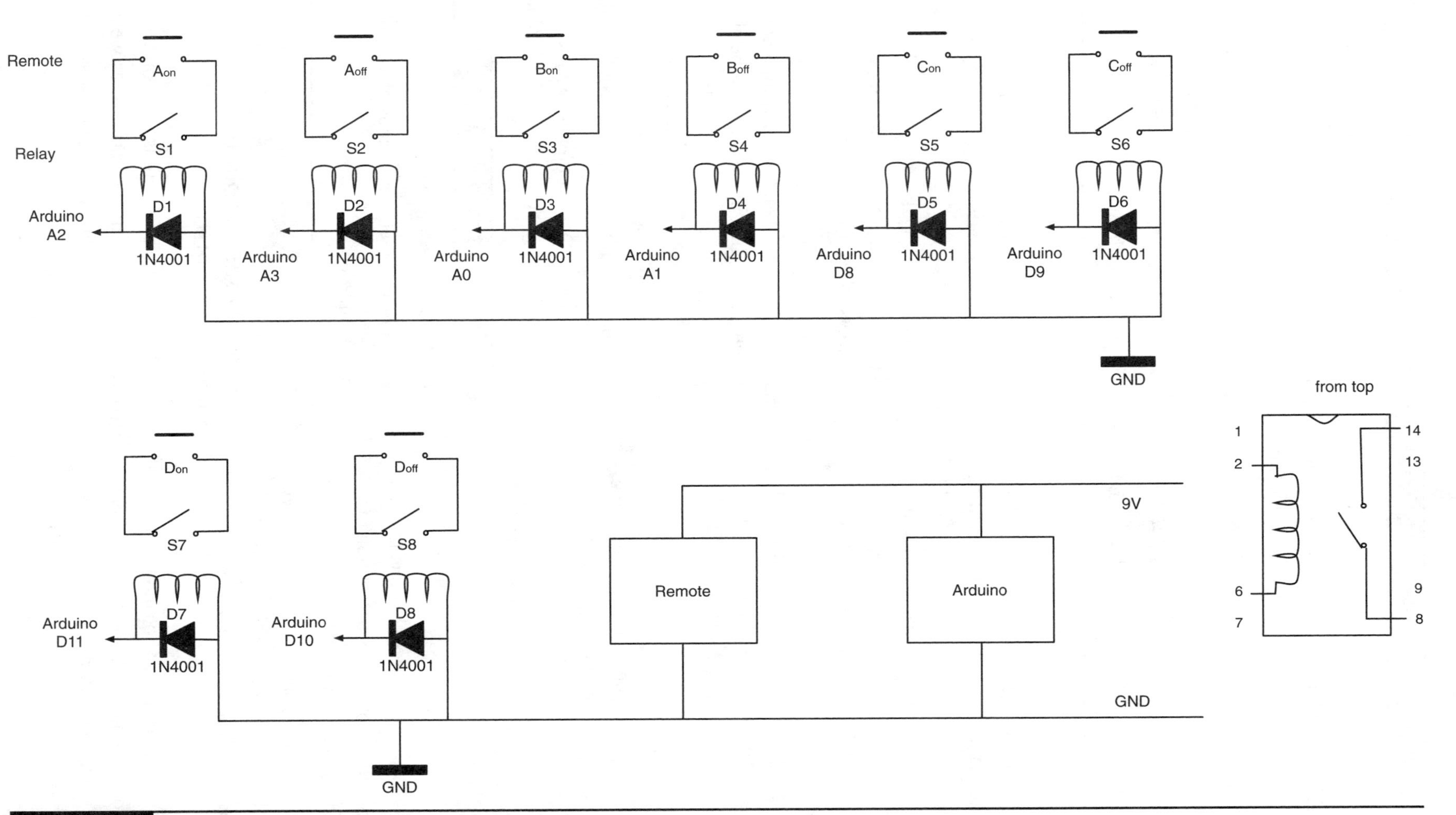

Figure 8-3 The schematic diagram for power control

PARTS BIN

Part	Quantity	Description	Source
Remote outlet kit	1	*See the following description*	eBay
Relay 1-8	8	5V SPST reed relay	eBay
D1-8	8	1N4001 diode	Farnell: 1458986
Perfboard	1		Farnell: 1172145
Header pins	1	Strip of 40 header pins	Farnell: 1097954
Nuts and bolts	4	Small nuts and bolts for fixing the perfboard	

There are a number of different remote outlets available on the market, and at the time of writing the construction will be simplest if you can find a device as similar as possible to that used by the author. This uses UK mains outlets and costs less than USD 10 for the remote and three remote control outlets. It was bought from a local supermarket. Other similar items are available in the U.S., including devices from La Crosse (RS-204U), Stanley (31164), and Woods (13568). Of these, the La Crosse is the closest to the device used by the author. But have a good look around because new devices are always coming on to the market. The system used by the author is shown in Figure 8-1.

The features to look for in a remote control power outlet set are:

- Essential, RF (Radio Frequency) NOT infrared.
- Essential, separate on and off buttons for each channel. Some remotes have one button that toggles. Do not use one of these.
- Desirable, four channels. If you can only get a device with three sets of on/off buttons, you can still build it, just leave out one set of relays.
- Desirable, dynamic outlet allocation. Some sets are hard-coded—that is, the button 1 will only switch the outlet with 1 printed on it. Others, such as the one the author used, allow you to "program" any outlet to be associated with any channel on the remote. This lets you set up groups of outlets all switched from the one channel, which adds a lot to the flexibility of the system. Indeed, you can buy extra sets of outlets to use from the one remote handset.

You are going to need eight reed relays. These can be expensive when bought from a conventional component supplier, but you should be able to find a lot of 10 for a few dollars on eBay. Look for ones that have the same pinout as that used by the author (see Figure 8-3). This should not be hard, as it is something of a standard pin configuration.

In addition to these components, you will also need the following tools listed in the Toolbox.

TOOLBOX

- Soldering equipment
- Multi-core wire in various colors
- A multimeter

Step 1. Disassemble the Remote Control

The first thing to point out is that you will most definitely be voiding the warranty, as we are going to be taking the remote apart and soldering leads to the circuit board.

The remote control used by the author is shown in Figures 8-4 and 8-5.

For the purposes of comparison, a different remote from another manufacturer is shown in Figures 8-6 and 8-7. The similarities between the two units are obvious.

Step 2. Attach Leads to the Remote Control PCB

Referring to Figure 8-4, what we are going to do is attach leads to each of the eight push switches on the main part of the board, as well as a lead for each of the two battery connections at the bottom of the figure. Note that we are not interested in the "All on" or "All off" buttons if the remote has such a feature.

If we turn the board over (Figure 8-5), we can see where the push buttons are soldered into place.

Each button actually has four connections, but we only need to find two which get connected when someone presses a button. This can be done using a short length of wire, since the remote will

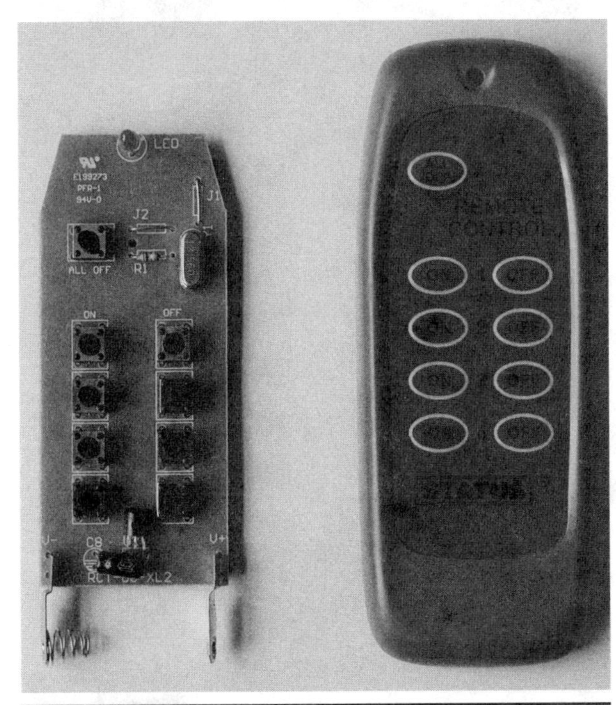

Figure 8-4 The remote from the top

Figure 8-5 The remote from underneath

Figure 8-6 The alternative remote from the front

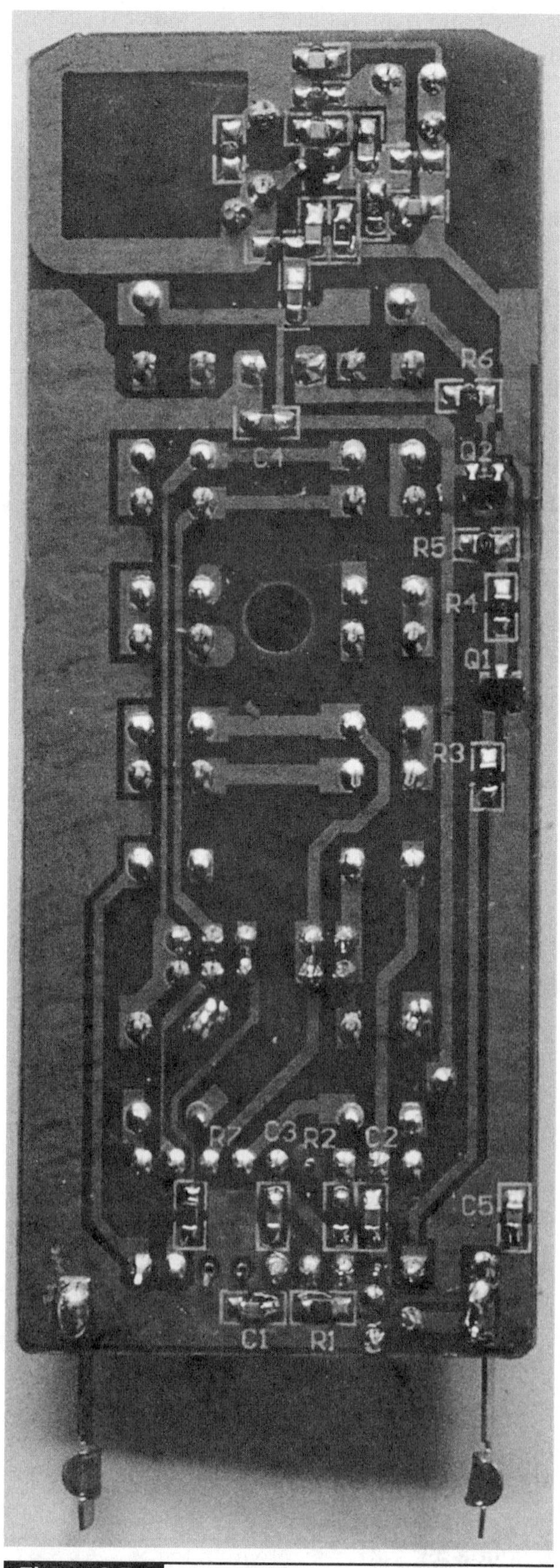

Figure 8-7 The alternative remote from underneath

have a little LED that comes on when a button is pressed. The procedure is as follows:

1. Fit the battery into the remote.
2. Identify the four connections for one of the switches.
3. Use the wire to short two of the connections for the switch, while watching to see if the LED comes on.
4. Make a note of the combination that works.
5. Repeat for each switch. You will probably find that it's the same pattern for every switch, but check.

Now cut lengths of solid core wire about half an inch (10mm) in length and solder them to all the contact points on the PCB, as shown in Figure 8-8. Note also the leads attached to the battery terminals.

Step 3. Place All the Components on the Perfboard

We are going to mount the remote control, the reed relays, and the Arduino onto a piece of perfboard. This material is rather like the stripboard we have used before, but without the strip. That is, it is just board drilled at 1/10-inch pitch. The idea is that you push components through from the top and solder them up underneath, connecting them together with wire. Figure 8-2 is what we are aiming for. Figure 8-9 shows the exact layout.

You may well have to adjust the spacing for your remote.

It is a bit fiddly getting all the wires on the remote to line up with holes in the perfboard. It can help to make them as straight as possible before trying to fit them onto the board, then use a screwdriver to nudge errant pins towards the hole. When all the pins are through the holes, bend some of them over just to ensure it stays in place.

Now we can place the reed relays, making sure they are the right way around (see the pinout on Figure 8-3). Afterward, place the diodes, again making sure they are the right way around.

The Arduino board is attached by pin headers. As you can see from Figure 8-10, not all of the four rows of connections on the Arduino are used.

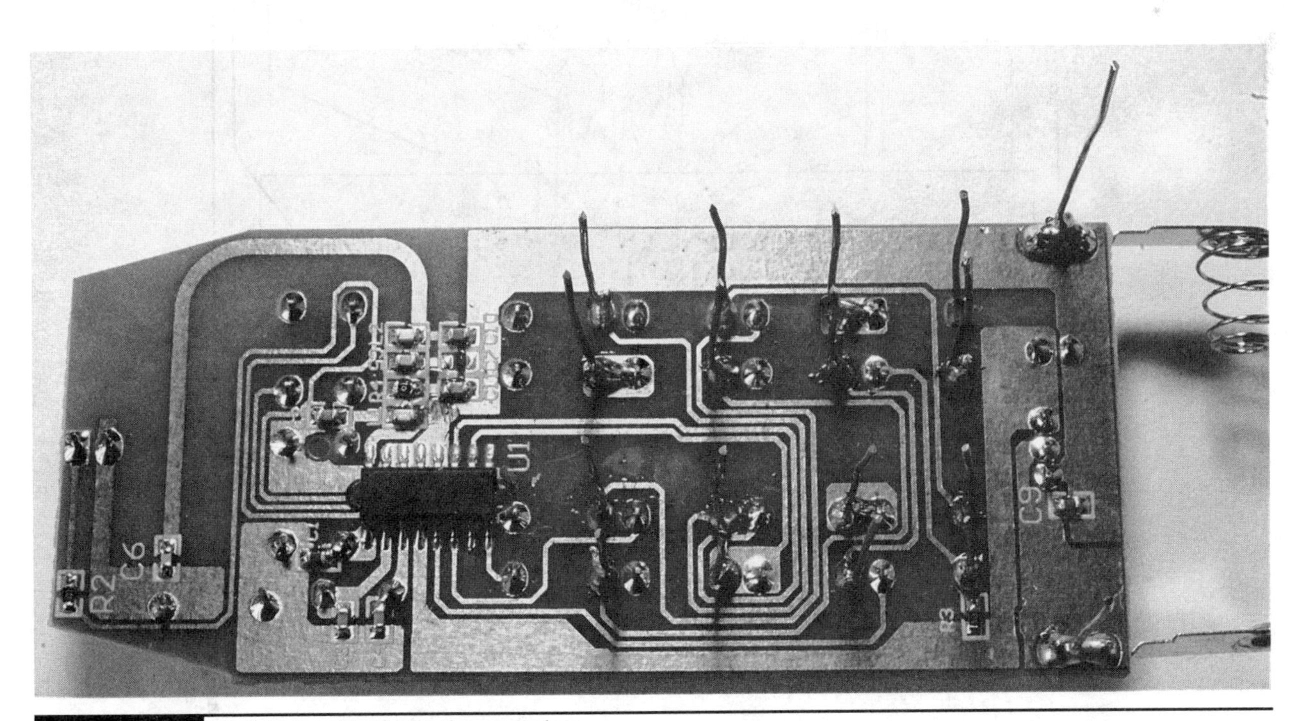

Figure 8-8 Soldering leads to the switches

View from the bottom of the board

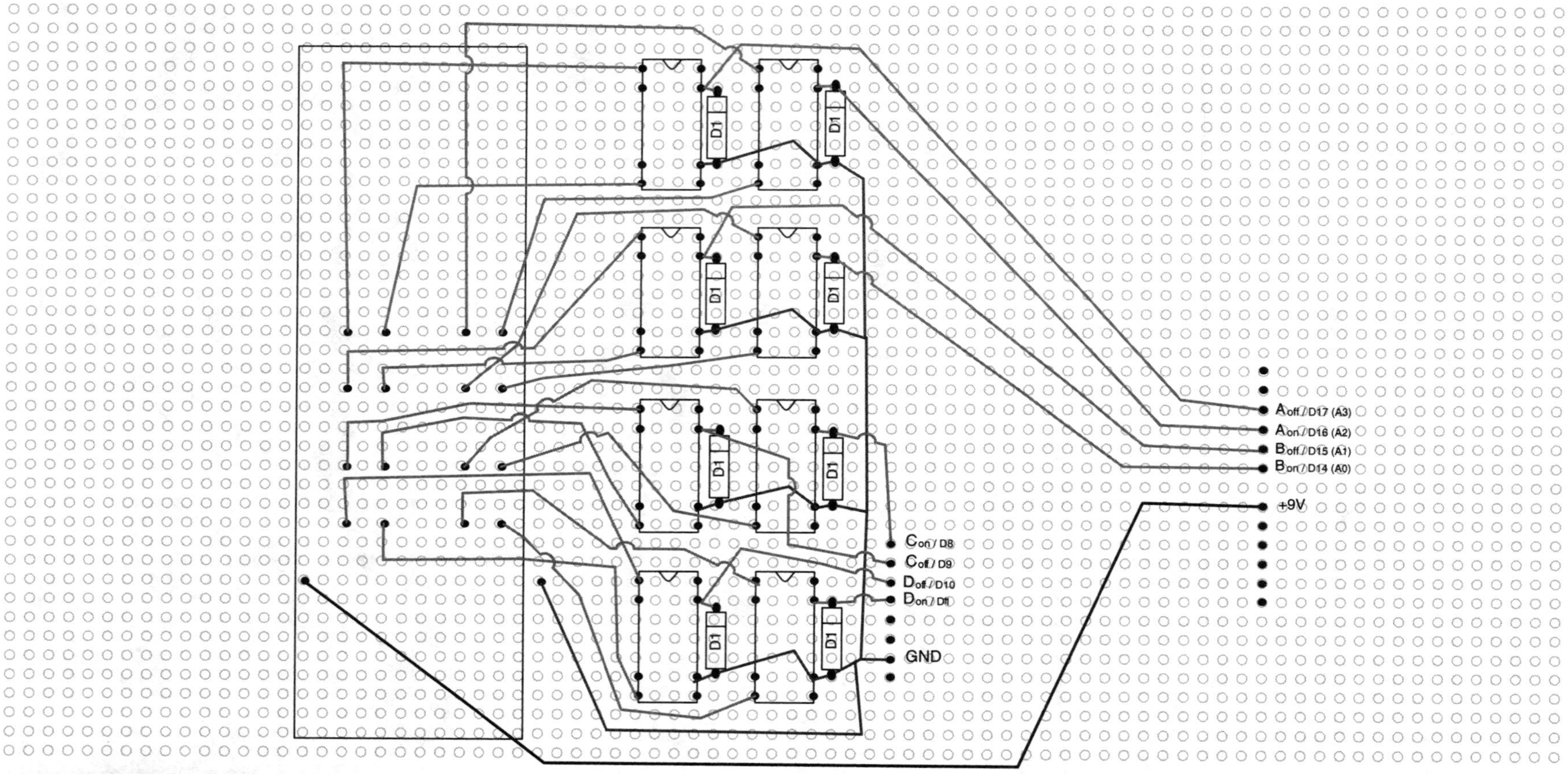

D1

Aoff / D17 (A3)
Aon / D16 (A2)
Boff / D15 (A1)
Bon / D14 (A0)
+9V

Con / D8
Coff / D9
Doff / D10
Don / D11
GND

Figure 8-9 The perfboard layout

There are two reasons for this. First, we do not need all the connections, and second, for some reason best known to the designers of the Arduino, the gap between the groups of pins on one side of the board (the left-hand side in Figure 8-10) do not fit a 0.1-inch pitch.

The Arduino is going to be attached using pin headers, so break three lengths off the header strip, one with eight pins and two with six pins each.

The best way to get the headers in the right position is to fit the headers into the sockets on the Arduino board (Figure 8-10) and then turn it over and fit it onto the perfboard.

Step 4. Solder the Connections

There are a lot of connections to make, and many wires cross over each other, so use insulated solid core wire for the connections. When all the connections are made, the back of your board should look like Figure 8-10.

It's not a bad idea to photocopy Figure 8-9 and check off each connection with a pen when it has been made.

When you think you have finished, go back through and check them all again.

Now, we are ready to test the board.

Step 5. Testing

To test our creation, we need a 9V power supply. If you do not have one, use the 15V power adapter and the 9V output from the sound interface you built in Chapter 7.

The web site www.duinodroid.com has a test sketch we can use (ch08_test_power). As we have already downloaded all the sketches, all that remains is to upload it onto your Arduino board. If you are not sure how to do this, please refer back to Chapter 7, Step 7.

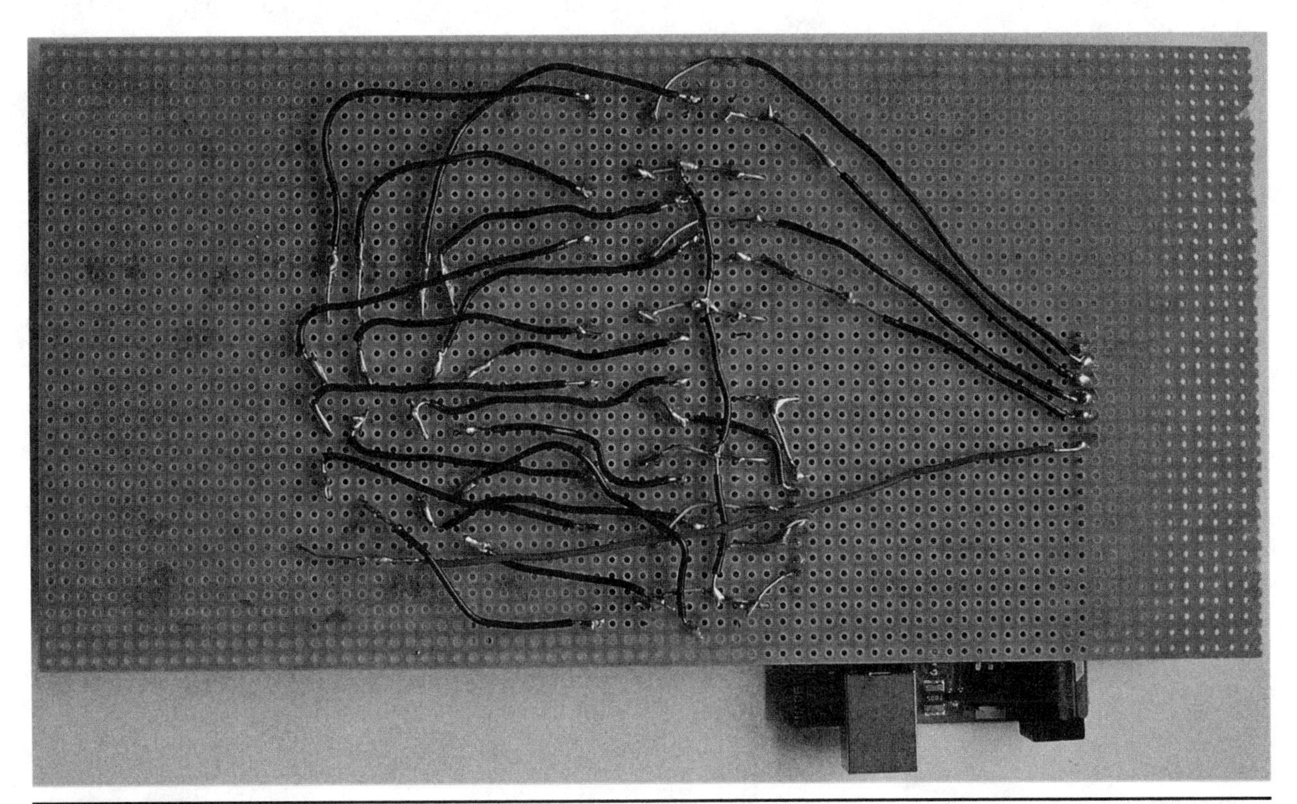

Figure 8-10 The back of the perfboard

Connect up the 9V supply to the Arduino, and from the Arduino software, launch the Serial Monitor (Figure 8-11).

The options shown by the sketch are the letters A to D in uppercase and lowercase. Typing an uppercase "A" into the Serial Monitor and pressing RETURN will send the command "A" to the Arduino board, which will effectively press the On button for channel A of the remote. Pressing lowercase "a" will turn it off.

To start with, just send each of the commands in turn and make sure the LED on the remote control lights up as if you had manually pressed the button. If this does not happen for one of the commands, check the wiring diagram of Figure 8-9 to try and find the problem with your board.

Now the exciting bit!

Plug one of the remote controlled outlets into a mains outlet and plug a table lamp into the remote outlet and turn it on. If your outlet is the type that needs programming, set it to the first channel on your remote. This is often done by pressing and holding the button on the front of the outlet until the LED flashes and then pressing a button on the remote.

Check that the remote still works normally by pressing the buttons on the front to turn the light on and off. Now we can try doing the same thing by sending the commands "A" and then "a" to the Arduino through the Serial Monitor.

Repeat this procedure for each of the channels on the remote to make sure that everything is working okay.

This is a useful project in its own right, as it allows you to control the power of devices remotely from your computer. However, our main use for this is to integrate it into our home automation system, which is exactly what we will do in the next section.

Adding It to the Home Automation Controller

We now know that our power control board works with the Arduino. It is time to take the next step and integrate it into the rest of the automation controller. To do this, we will attach the sound interface board from Chapter 7 to the perfboard, and make the connections from the sound card to the Arduino +5V and A4 pins. Figure 8-12 shows the revised wiring diagram for the project.

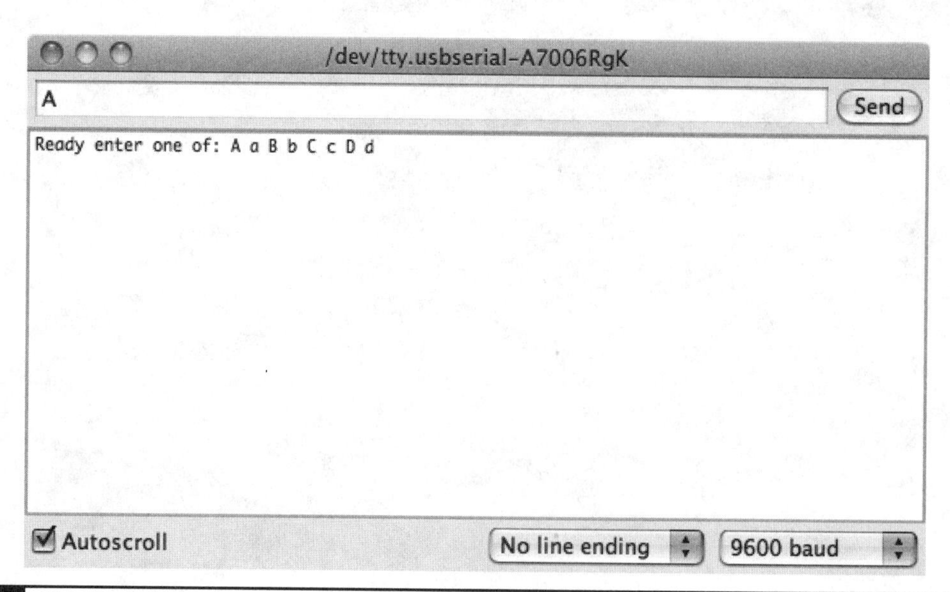

Figure 8-11 The Serial Monitor for power test sketch

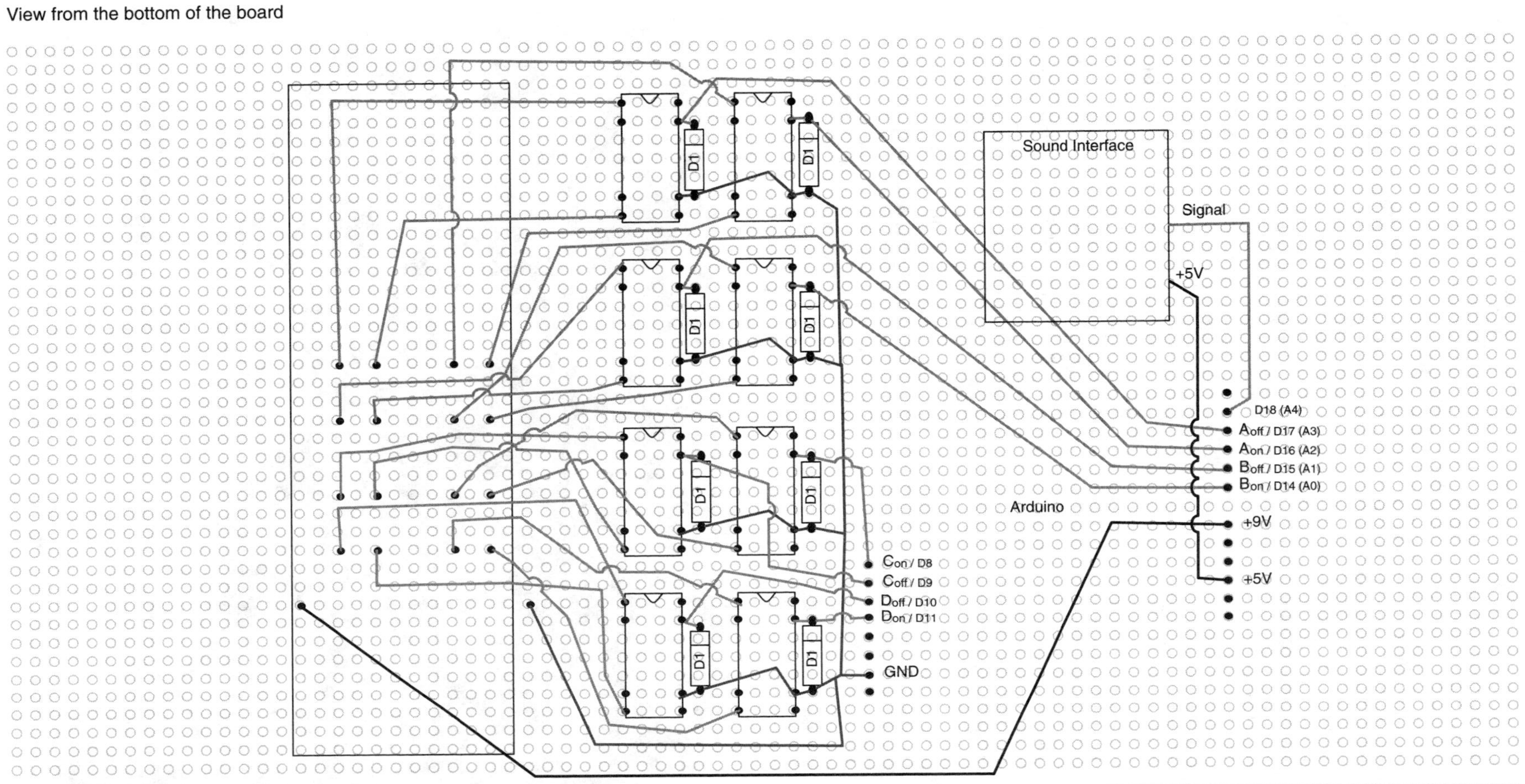

Figure 8-12 The wiring diagram with the sound interface

Figure 8-13 shows the board from the top with the sound interface attached. Notice how the Arduino is now receiving its power from the 9V flying lead and plug from the sound interface, and the 3.5mm audio plug is ready to be plugged into the headphone output of the Android tablet.

Before we fit everything into the box. Let's try the system out on the bench. To do this, we will use the full "Home Automation" Android application, but employ the "ch07_sound_test" sketch, which we have just used on the Arduino board. This will let us check that the messages we are sending from the Android tablet are being received correctly, before we change over to using the proper Arduino sketch. So upload the "ch07_sound_test" to the Arduino board.

Make sure everything is connected up, that is:

- Connect the 15V supply to the sound interface board.
- Connect the 9V plug from the sound board to the Arduino.
- Plug the 3.5mm headphone plug into the Android tablet.
- Plug the Arduino board into your computer and open the Serial Monitor.
- Start up the Home Automation app on the Android tablet and make sure the sound is turned up.

Open the browser on the Home Automation app and click the "Light and Power Outlets" button. Now press the button to turn on Outlet 1. The Serial Monitor should display the number 0101. Click "Off" for that outlet. It should say 0100.

Now we can replace the test sketch for the Arduino with the real one. As usual, this can be downloaded from www.duinodroid.com with the other sketches for the book. The sketch is called "ch07_sound_test."

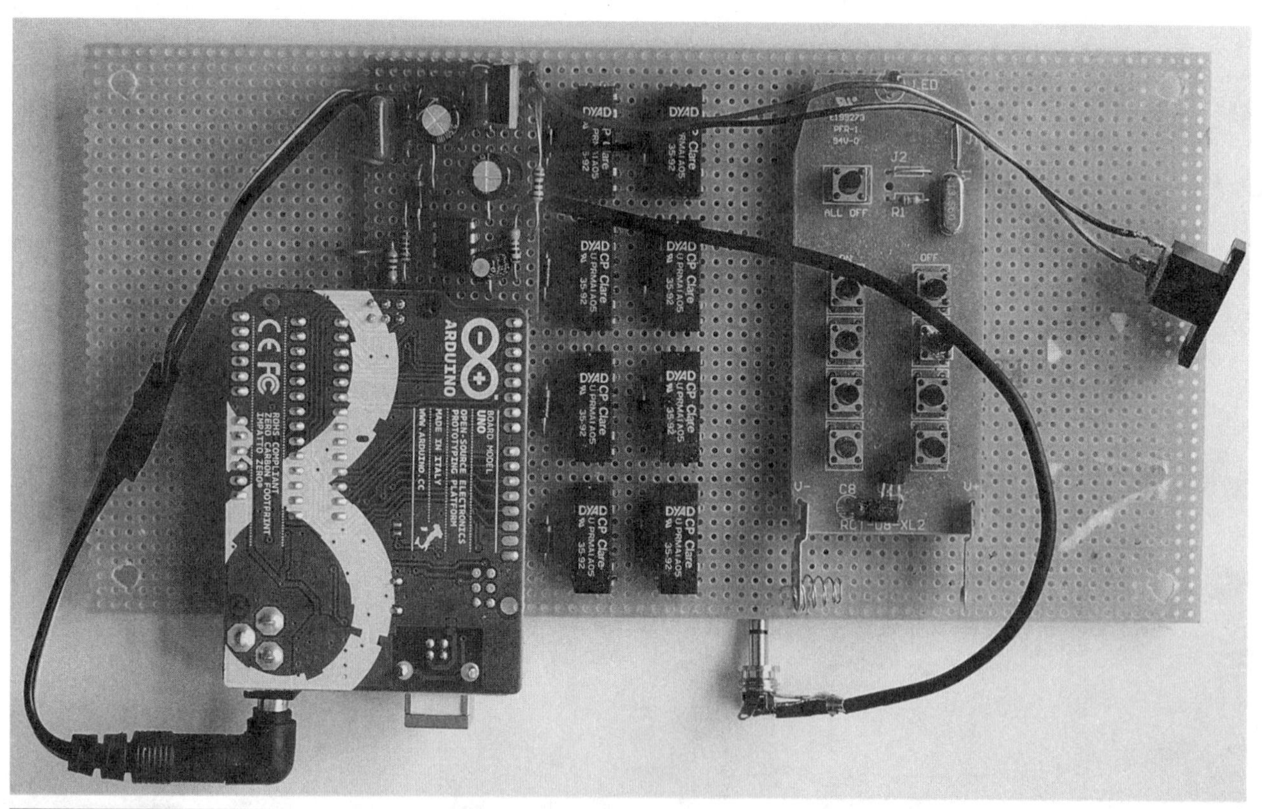

Figure 8-13 The power control board with the sound interface attached

The easiest way to test the system is to attach a table lamp to one of the remote outlets, run the Home Automation app on the Android tablet, launch the browser, and go to the "Lights and Power Outlets" page. If the outlet is of the type that must be programmed to respond to a particular control, then we need to do this instead: Press and hold the button on the front of the outlet until the LED flashes and then press the "On" button in the browser you want to associate with that outlet. I suggest using the first outlet. The outlet LED should stop blinking to indicate it has been programmed.

If you are using the type of remote outlet system with preprogrammed numbered outlets, then you do not need to carry out the preceding step.

You should now find that you can turn the lamp on and off from the Android tablet. I will leave you alone now to try out all the various features of the home automation controller, and perhaps try connecting to it through another computer on your network, or even from the Internet (see Chapter 7).

Once you have recovered from the excitement, you need to fit everything neatly into the project box.

We already have the basis for a boxed project following Chapter 7; however, we now need to accommodate our power control perfboard. So, some more holes must be drilled into the project box. We need holes for four screws to attach the perfboard, a hole for the 15V power socket attached by flying lead to the sound card, and optionally a hole big enough for a USB cable to be attached to the Arduino board. This last hole is only really necessary if you plan to modify your software on the Arduino.

The perfboard should be secured to the base of the project box with nuts and bolts (Figure 8-14). You also need to fit and re-solder the power socket.

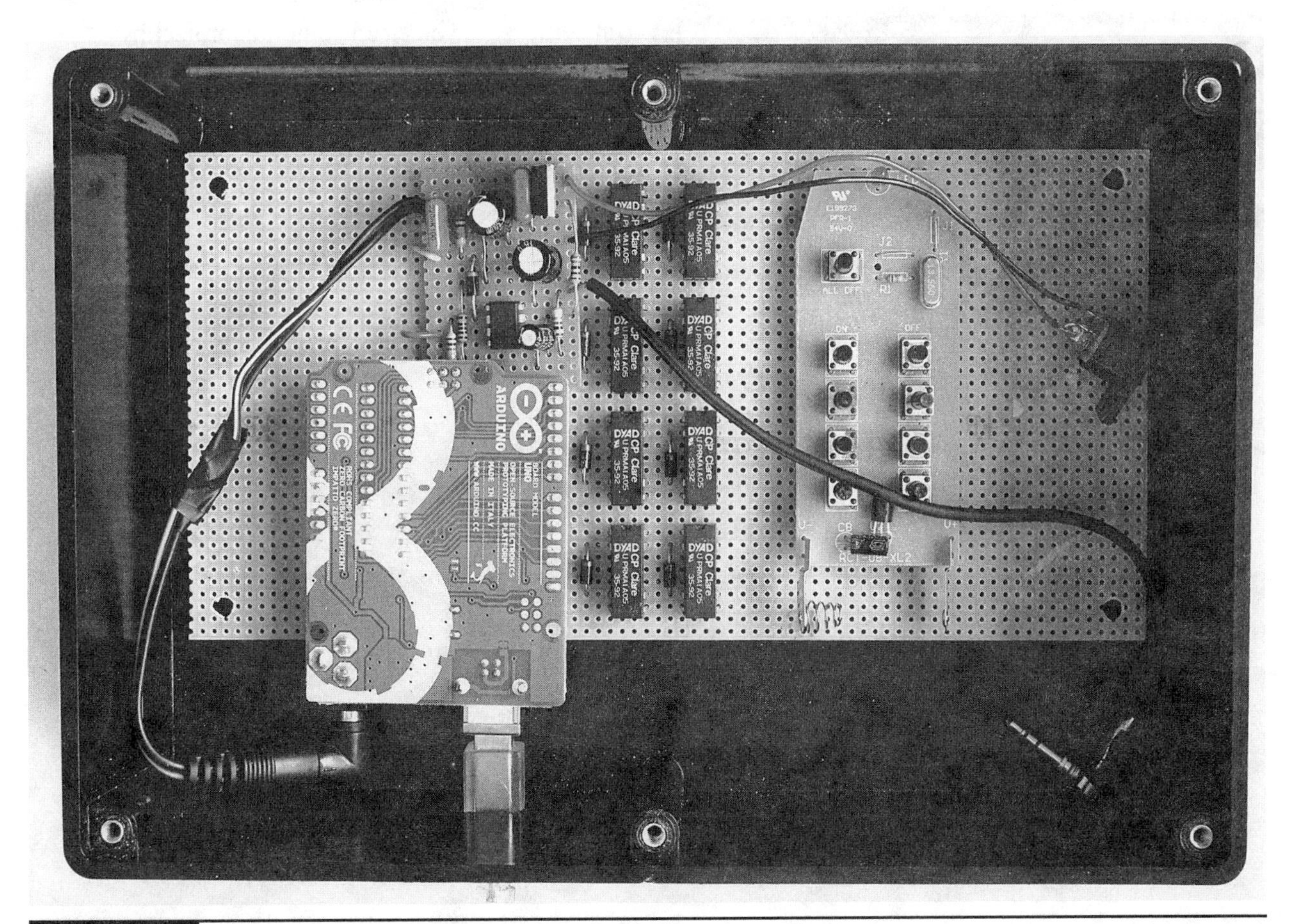

Figure 8-14 The inside of the project box with perfboard attached

Setting Up Your Home

Connecting the power outlets is easy enough, and if you have the kind that are programmable, you will probably want to associate a group of power outlets with one channel on the home automation controller. For instance, using a series of remote control outlets, set up all the appliances you want to turn off at night.

Figure 8-15 shows the arrangement of outlets used by the author. All the downstairs lights are grouped onto one channel—handy for bed time.

The main audio visual and TV appliances in the living room are grouped onto another channel, and the minions' sleeping quarters onto the remaining two channels.

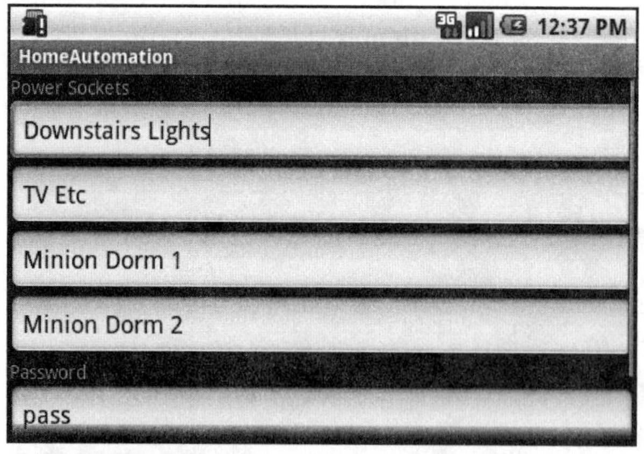

Figure 8-15 An example setup for home automation

As well as outlets, it is also really useful to be able to control ceiling and wall lights. This is more difficult than the wall outlets. The most direct approach to this unfortunately requires some expertise in wiring. It should therefore only be carried out by a suitably qualified electrician.

You can convert a remote control power socket to control lights using a screw terminal block (see Figure 8-16). Do not take this step unless you are suitably qualified to alter your domestic wiring.

Theory

In this chapter, we will examine the software for our project, in particular the two Arduino sketches we used, "ch08_test_power" and "ch08_home_automation." But first we will look at some of the electronics in this project by examining the relays we used to simulate button presses on the remote control.

Relays

Relays are, in electronics terms, an ancient technology. They were around before vacuum tubes, and were used as switches to control power

Figure 8-16 Converting a remote control outlet for lighting

and operate telephone exchanges. For many applications, they have been replaced by transistors and ICs. However, they still have a place in the high-tech world.

The type of relays used in this project are called reed relays (Figure 8-17). They are no good for switching high currents, but they are fine for this project. They also have the great advantage that, unlike most relays, they can be controlled directly by an Arduino board without needing a transistor.

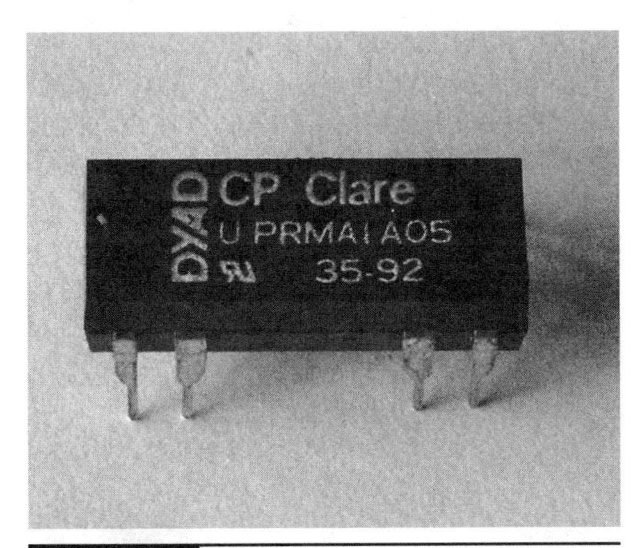

Figure 8-17 A reed relay

So, what exactly is a relay? Well, it has two parts that are electrically isolated from each other. There is a coil that acts as an electromagnet and a switch that is activated by the electromagnet. In the case of the reed relay, the switch is made by two thin contact reeds in a small sealed glass envelope. This makes them very reliable and long-lived.

Test Sketch

Here is the code for the Power Test sketch we used earlier.

```
// Power Control test sketch

#define AonPin 16
#define AoffPin 17
#define BonPin 14
#define BoffPin 15
#define ConPin 8
#define CoffPin 9
#define DonPin 11
#define DoffPin 10

#define ButtonPressPeriod 500

int onPins[] = {AonPin, BonPin, ConPin,
    DonPin};
int offPins[] = {AoffPin, BoffPin,
    CoffPin, DoffPin};

void setup()
{
  for (int i = 0; i < 4; i++)
  {
    pinMode(onPins[i], OUTPUT);
    pinMode(offPins[i], OUTPUT);
  }
  Serial.begin(9600);
  Serial.println("Ready enter one of:
   A a B b C c D d");
}

void loop()
{
  int channel = 0;
  if (Serial.available())
  {
    char ch = Serial.read();
    if (ch >= 'a' && ch <= 'd')
    {
      channel = ch - 'a';
      pressButton(channel, offPins);
    }
    else if (ch >= 'A' && ch <= 'D')
    {
      channel = ch - 'A';
      pressButton(channel, onPins);
    }
  }
}

void pressButton(int channel, int
    column[])
{
  digitalWrite(column[channel], HIGH);
  delay(ButtonPressPeriod);
  digitalWrite(column[channel], LOW);
}
```

The first thing we do is use #define commands to specify all the Arduino pins we will use. Note that we are using the analog pins of the Arduino as digital pins. This is possible with the Arduino software, which allows you to address the analog pins A0 to A5 as if they were digital pins by adding 14 to their pin number. So 16 becomes A2, 17 becomes A3, and so on.

We then define another constant—ButtonPressPeriod—that specifies the time in milliseconds for which each relay should be closed to simulate a button press.

To make it easier to access the pins concerned with the relays, we group them into two arrays, one for the "on" pins and one for the "off" pins.

The "setup" function makes all those pins outputs, as well as initializes the serial connection back to the Serial Monitor.

As usual, most of the action takes place in the main "loop" function. This checks to see if a command has been sent over the Serial Monitor (Serial.available) and if one has, and it's between "a" and "d" or "A" and "D", it works out the channel number between 0 and 4 and then calls the function "pressButton" with the channel as one argument, and the "turning on" array or "turning off" array as appropriate.

The "pressButton" function then turns on the appropriate output pin, which will in turn activate the relevant relay to press the button and turn on the channel.

Real Sketch

The real sketch for the home automation controller is a combination of the sound test sketch from Chapter 7 and the power test sketch we have just looked at.

```
#include <VirtualWire.h>

#define soundPin 18
#define zeroDurationFrom 10000
#define zeroDurationTo 25000
#define oneDurationFrom 35000
#define oneDurationTo 50000
#define resetTimeout 3000

#define AonPin 16
#define AoffPin 17
#define BonPin 14
#define BoffPin 15
#define ConPin 8
#define CoffPin 9
#define DonPin 11
#define DoffPin 10

#define ButtonPressPeriod 1000

int onPins[] = {AonPin, BonPin, ConPin,
   DonPin};
int offPins[] = {AoffPin, BoffPin,
   CoffPin, DoffPin};
int remote = 0;

void setup()
{
  pinMode(soundPin, INPUT);
  for (int i = 0; i < 4; i++)
  {
    pinMode(onPins[i], OUTPUT);
    pinMode(offPins[i], OUTPUT);
  }
  vw_set_ptt_pin(5); // out of the way
  vw_set_rx_pin(4); // out of the way
  vw_set_ptt_inverted(true);
  vw_setup(2000);
  Serial.begin(9600);
}

unsigned int result;
int bitNo = 0;
long lastPulseTime = 0;

void loop()
{
  long pulseLength = pulseIn(soundPin,
   HIGH, oneDurationTo * 2);
```

```
  long timeSinceLastPulse = millis() -
   lastPulseTime;
  lastPulseTime = millis();
  if (pulseLength == 0 ||
   timeSinceLastPulse > resetTimeout)
  {
    bitNo = 0; result = 0;
  }
  else
  {
    if (pulseLength >= zeroDurationFrom
    && pulseLength <= zeroDurationTo)
    {
      result = result << 1;
      bitNo ++;
    }
    else if (pulseLength >=
    oneDurationFrom && pulseLength
    <= 50000)
    {
      result = (result << 1) + 1;
      bitNo ++;
    }
  }
  if (bitNo == 16)
  {
    processWord(result);
    bitNo = 0; result = 0;
  }
}

void processWord(int message)
{
    int device = message >> 8;
    int action = message & 0x00FF;
    Serial.print("Device: ");
   Serial.print(device);
    Serial.print(" action: ");
   Serial.println(action);
    if (device > 0 && device <= 4)
    {
      processPower(device, action);
    }
    else
    {
      processRadio(device, action);
    }
}

void processPower(int device, int
   action)
{
  if (action)
  {
    pressButton(device, onPins);
  }
  else
  {
    pressButton(device, offPins);
  }
}

void processRadio(int device, int
   action)
{
    uint8_t msg[2];
    msg[0] = (uint8_t)device;
    msg[1] = (uint8_t)action;

    vw_send((uint8_t *)msg, 2);
    delay(400);
}

void pressButton(int channel, int
   column[])
{
  int pin = column[channel - 1];
  digitalWrite(column[channel-1],
   HIGH);
  delay(ButtonPressPeriod);
  digitalWrite(column[channel-1], LOW);
  delay(ButtonPressPeriod);
}
```

As with the previous sketch, we start by using #define to declare some constant values, such as the pins we will use for the relays, as well as the durations for the pulses coming from the sound interface (see the "Theory" section of Chapter 7).

In this case, once all 16 bits have been gathered, they are passed as the parameter result to the function "processWord". This splits them into "device" and "action". If the device is in the range 1 to 4, then we know it is one of the power channels, so we call the same "pressButton" function we did in the previous sketch.

This sketch also includes code to handle the door and the heating thermostat—in fact, both are handled by the "processRadio" function and thus can be disregarded for now.

Summary

Our home automation controller is now a useful device. We can turn things on and off from anywhere, as well as set times. In the next chapter, we will extend this design a bit further by adding a link to a heating thermostat.

CHAPTER 9

Smart Thermostat

Minions are notoriously difficult to keep at the right temperature. Too cold and they become hypothermic, and just sit there looking blue and are incapable of carrying out their master's evil schemes. On the other hand, if they get too hot, they tend to perspire heavily and drip all over the Evil Genius' plans for global domination.

This project provides the last word in temperature regulation. It not only acts as a stand-alone thermostat, but also has a RF link from the Home Automation controller, that lets you set different temperatures for different times of day.

The home automation will tell the thermostat its new set temperature regularly every 10 seconds over the wireless link. To give us some feedback that everything is working, the LED will blink twice each time it receives a message.

To keep the project small and reduce costs, the design uses the microcontroller from an Arduino, but takes it off the Arduino onto a piece of stripboard. The Arduino is used to program the microcontroller, but then is replaced with another microcontroller chip. An ATMega328 used in an Arduino is readily available and costs around USD 5.

Figure 9-1 shows the thermostat, which just has an override switch, a control knob, and an LED.

Most of the software for the home thermostat belongs with the Home Automation controller, which provides a nice touchscreen graphical

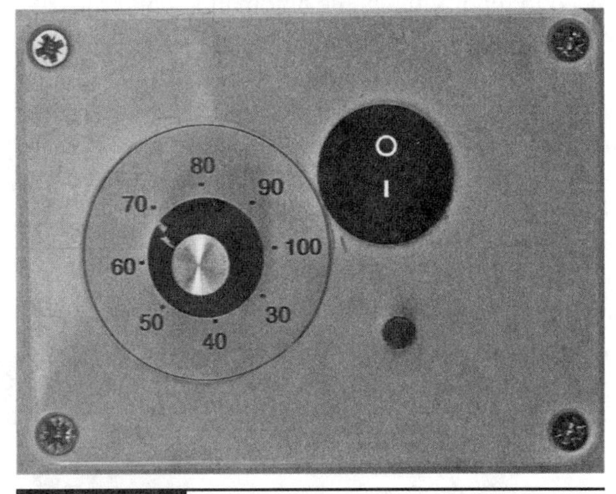

Figure 9-1 A smart heating thermostat

mechanism for setting the temperature profile (Figure 9-2). There are two pages for setting the temperature profile. One for weekdays, and one for weekends. The day is divided up into time slots.

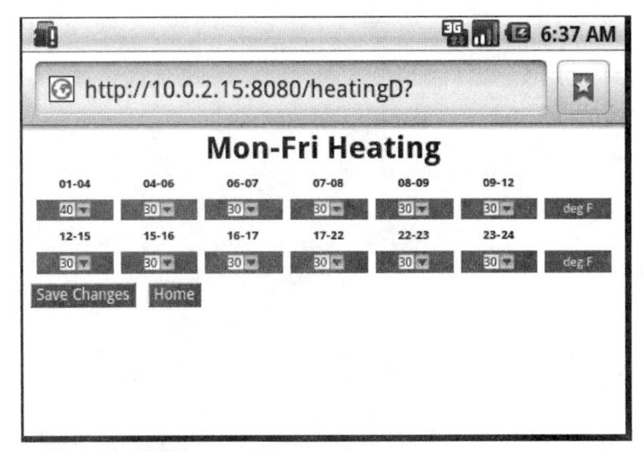

Figure 9-2 The home automation controller—temperature control

So, 01-04 refers to 1 a.m. to 4 a.m. The temperature for each time slot is set from a dropdown list.

If you flip the thermostat's "override" switch, it stops using the temperatures being sent from the home automation controller and just uses the temperature set on the control dial.

Construction

We need to build the thermostat itself, but we also need to modify the home automation controller, providing it with the RF transmitter and updating the software on its Arduino.

It is best to make the thermostat first, so that we can use it to test that the modifications to the home automation controller were successful.

Apart from the microcontroller IC and a few supporting components, there is just a transistor, a relay, and a RF receiver module on the stripboard. We use a reclaimed 5V power supply from a mobile phone charger. If you do not have such a thing, then a normal "wall-wart" power supply can be used.

Figure 9-3 shows the relationship between the thermostat, the central heating system, and the home automation controller.

Figure 9-4 shows the schematic diagram for the thermostat.

CAUTION **This project switches high-voltage domestic electricity and involves modifying the wiring to your home heating system. Domestic household electricity kills many people each year and starts hundreds of fires in the U.S. alone.**

You should only install this system if you are suitably qualified to do so and have a good understanding of the safety implications of altering your home's wiring.

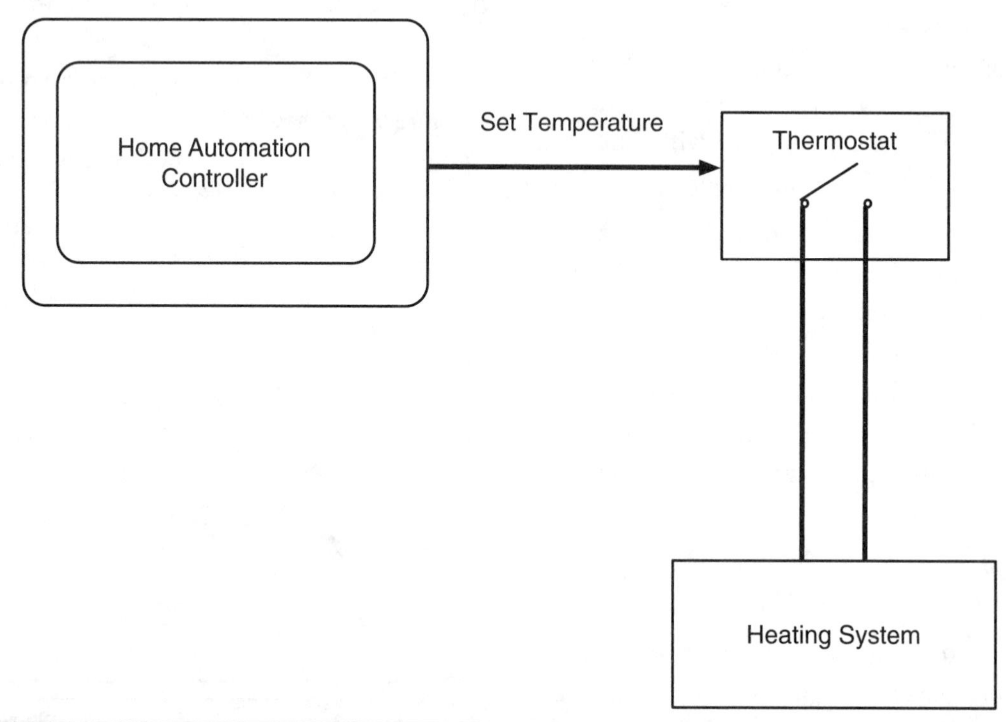

Figure 9-3 The home automation controller and thermostat

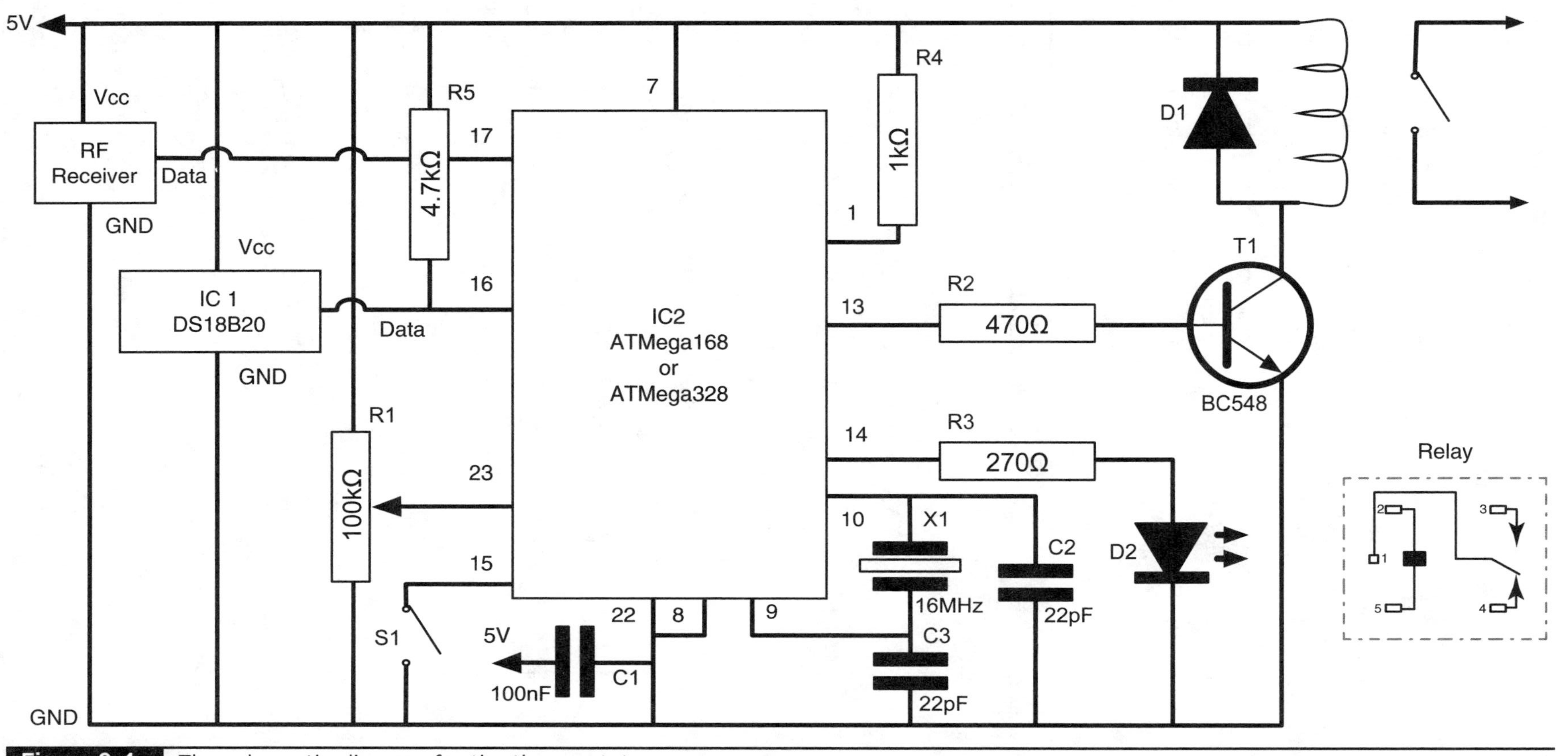

Figure 9-4 The schematic diagram for the thermostat

What You Will Need

You will need the following components in the Parts Bin to make the thermostat.

You will also need an Arduino Uno or Duemilanove to program the ATMega328 or ATMega168 chip. When you buy this, make sure you get a version with the Arduino bootloader already installed on it.

There are many 433-MHz RF modules available on the market, and most have the same simple-to-use pinout. But check the datasheets for the modules you intend to buy against Figure 9-4 to make sure they have the same pin connections.

The choice of relay depends on which country you live in. It will need to have a 5 or 6V coil. The contacts must be of a similar specification to the thermostat you are replacing. If you are not sure what the current and voltage requirements of the relay contacts are, consult a heating engineer. Please note that the relay specified was suitable for the author's central heating system. The author has no way of knowing if this is suitable for your own heating system, so do the research.

PARTS BIN

Part	Quantity	Description	Source
Microcontroller	1	ATMega328 with bootloader	Sparkfun: DEV-10524
RF receiver	1	433-MHz receiver module	Farnell: 1304026
RF transmitter	1	433-MHz transmitter module	Farnell: 1304024
Relay	1	5 or 6V relay	Farnell: 1455502
R1	1	100kΩ variable resistor	
Knob	1	Knob to fit R1	Farnell: 1282303
R2	1	470Ω 0.5W metal film resistor	Farnell: 1099883
R3	1	270Ω 0.5W metal film resistor	Farnell: 9340300
R4	1	1kΩ 0.5W metal film resistor	Farnell: 9339779
R5	1	4.7kΩ 0.5W metal film resistor	Farnell: 9340629
C1	1	100nF ceramic capacitor	Farnell: 1200414
C2, C3	2	22pF ceramic capacitor	Farnell: 1600966
X1	1	16-MHz crystal	Farnell: 1611761
D1	1	1N4001	Farnell: 1458986
D2	1	5mm red LED	Farnell: 1712786
IC1	1	DS18B20	Sparkfun: SEN-00245
T1	1	BC548	Farnell: 1467872
S1	1	Plastic rocker switch	Farnell: 1634645
IC socket	1	28-pin DIP IC socket	Farnell: 1824463
Power supply	1	5V power supply	*See description*
Terminal strip	1	2-way 2A screw terminal strip	Hardware store
Stripboard	1	29 strips of 24 holes	Farnell: 1201473
Box	1	Small project box	RS/local electronics store

The minions go through cell phones like there's no tomorrow. So there are always surplus chargers in the lair, ancient relics of long superseded phones. Of course, the Evil Genius likes to do his bit to save the planet—just before conquering it.

To this end, we are going to reuse one of these chargers to provide the 5V power to the thermostat. If you do not have one, you will have to buy a 5V wall-wart style power supply.

In addition to these components, you will also need the tools listed in the Toolbox.

TOOLBOX

- Soldering equipment
- Craft knife and ruler
- Hot glue gun or epoxy glue
- Multi-core wire in various colors
- A multimeter

Step 1. Prepare the Stripboard

Figure 9-5 shows the stripboard layout for the thermostat.

Cut the board to the right size. The neatest way to do this is to score the board heavily with a craft knife and then break it over the edge of a table. You can also use a strong pair of scissors, but the result will not be as neat.

As you can see, there are quite a few cuts to be made in the board. As with the sound interface in Chapter 1, use a drill bit to make breaks in the track where indicated. Note that the board is shown from the top in Figure 9-5.

When all 19 breaks have been made, your board should look something like Figure 9-6.

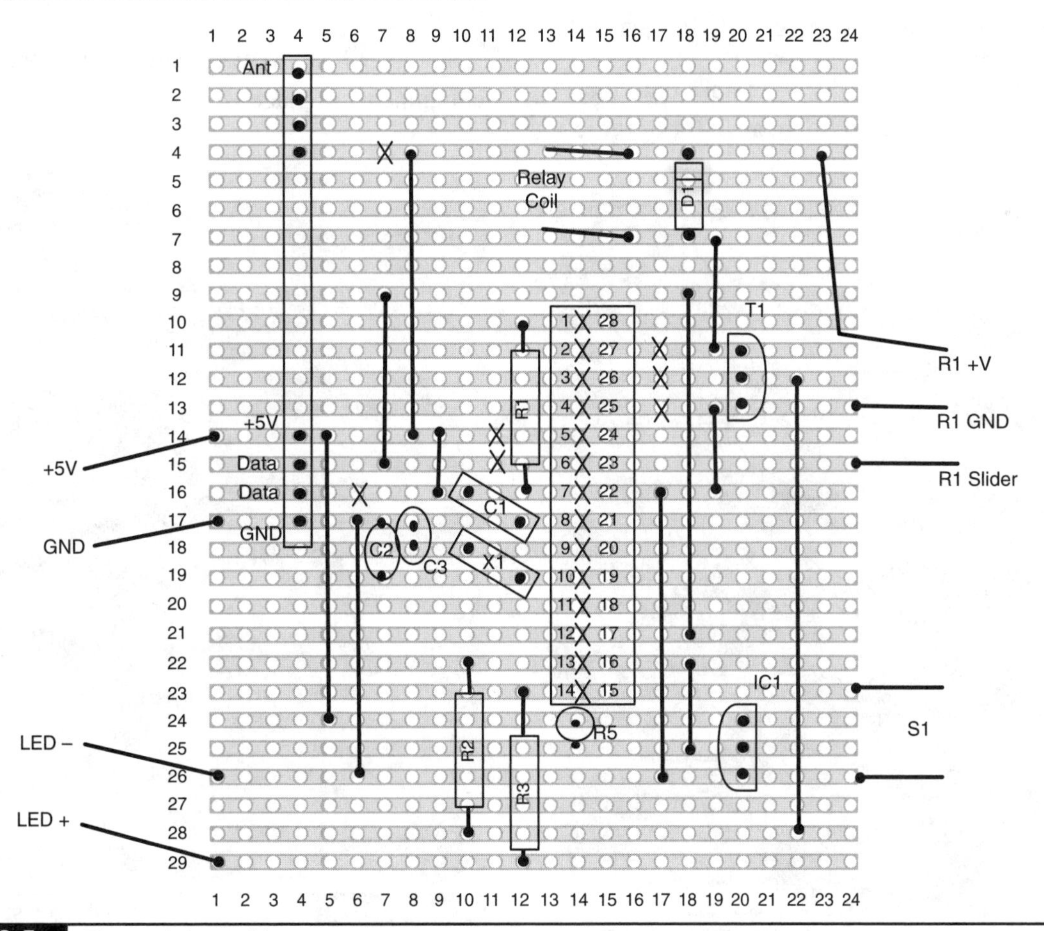

Figure 9-5 The stripboard layout

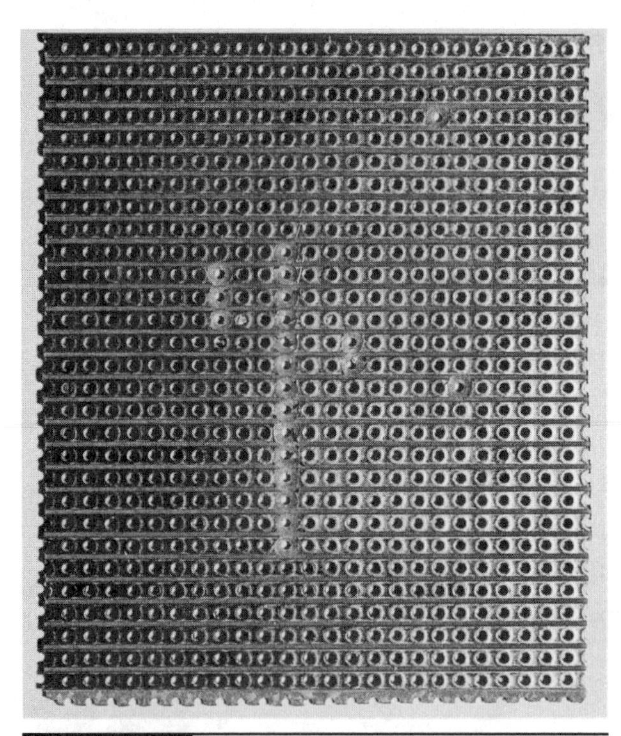

Figure 9-6 The prepared stripboard

Step 2. Solder the Links and IC Socket

The next step is to solder the wire links onto the board (Figure 9-7). Again, there are a lot of these (11), so take care that you have made all of them and got them all in the right places.

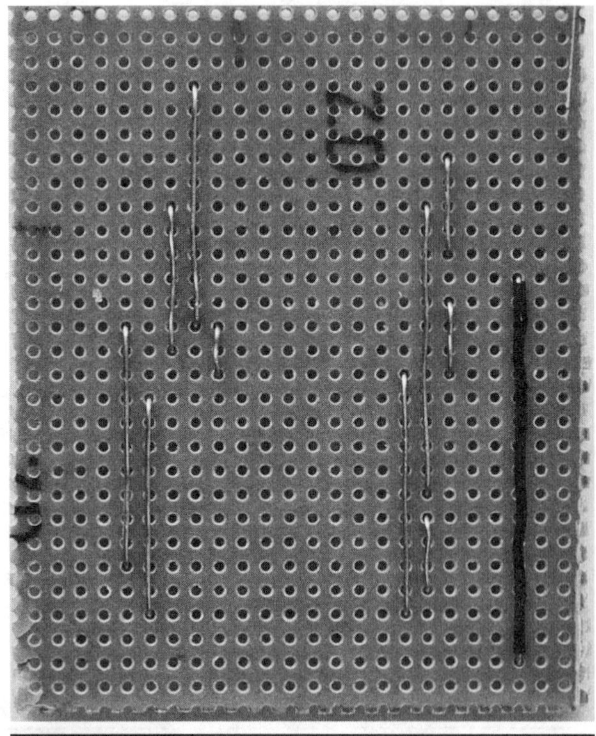

Figure 9-7 The stripboard with links

Step 3. Solder the Resistors and Diode

We can now start on the real components. As always, begin with the lowest profile components first. That means the resistors and the diode, then solder in the capacitors. When soldering the diode, make sure it is the right way around. The striped end should be toward the bottom of the board.

R5 is mounted vertically, so leave it until the next step when we solder in the capacitors and higher components.

Once they are all soldered into place, you can solder the IC socket. It is a good idea to arrange the socket so it has the little cutout marker to indicate the end with pin 1 toward the top of the board.

Figure 9-8 shows the board with all these components in place.

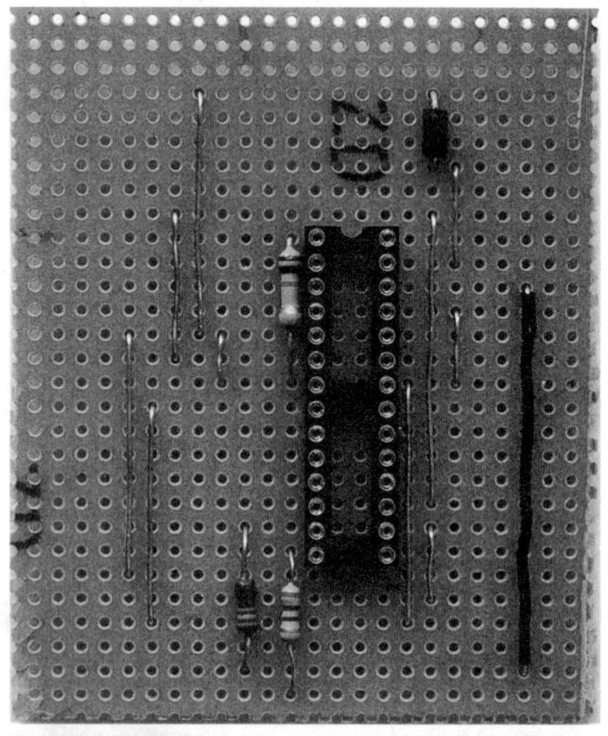

Figure 9-8 The stripboard with resistors and capacitors

Step 4. Solder the Remaining Components

The remaining components can now be attached. Take care with IC1 and T1 to ensure they are both the right way around and that you do not heat the leads for too long.

We only use the bottom four pins of the RF receiver module; however, if your thermostat is a long way from the home automation controller, you may need to attach a wire to the ANT connection to act as an antenna. The wire should be about 100mm long.

When everything is connected up, it should look something like Figure 9-9.

At this point, a thorough inspection of the board is a good idea. Inspect the top of the board to make sure all the components and links are in the places indicated in Figure 9-5, and check the bottom of the board to make sure all the breaks are in the right place and that there are no accidental bridges of solder between tracks.

Figure 9-9 The completed stripboard

Step 5. Program and Install the Microcontroller

We need to program the microcontroller with the Arduino sketch to control the thermostat. The easiest way to do this is to use an Arduino board.

It usually makes sense to program the microcontroller IC that's already in the Arduino board and replace it with the chip you bought, once the chip has been transferred to the stripboard.

The sketch uses three libraries, two for the temperature sensor (OneWire and DallasTemperature) and the VirtualWire library for the RF receiver. You have probably already installed the VirtualWire library in Chapter 8. You will, however, have to install the other two libraries into your Arduino environment on your computer.

The procedure is the same as for all libraries (see Chapter 1), and you can download the OneWire and Dallas libraries from www.hacktronics.com/code/OneWire-v2.zip.

So, just install the sketch (ch09_thermostat) onto the Arduino board. Then, with the power disconnected, carefully remove the IC from the board and fit it into the socket on the stripboard, making sure it's the right way around.

When removing the IC, be very careful because it is easy to bend or break off the pins. Ease the IC out at each end of the socket using a knife, a bit at a time until it comes loose. You should also observe anti-static precautions, at least touching grounded metal to discharge yourself before touching the device's pins. An earthing strap is another option.

Step 6. Connect Everything Up

The board is now complete, so after a thorough inspection to make sure there are no unwanted solder bridges between tracks, we can connect everything up (Figure 9-10). We will not fit it into its box until we are sure everything is working.

We are using a redundant mobile charger as the power supply. We do not need a powerful charger—as long as it is 5V and can supply at least 100mA, it will be fine. With the charger unplugged, cut off the plug at the end of the lead and strip the wires so you can attach a multimeter

Figure 9-10 Connecting up the board

and test the voltage. First, make sure it is indeed a stable 5V, and second, determine which is the positive lead. For now, the wires can be soldered directly to the stripboard.

Next, we attach the leads to the variable resistor. Note that the order of the leads to the variable on the stripboard are not in the order you might expect. Looking at the back of the variable resistor, with the leads downwards, the positive lead will be on the left, the negative on the right, and the slider in the middle.

The switch and LED should both be attached using short lengths of multi-core wire. It is a good idea to use color-coded leads for the LED, so as to prevent accidentally getting the polarity wrong.

The leads to the relay coil can be connected either way around. Glue the relay to the board and attach the coil leads. Solder the leads to the normally open relay contacts and attach them to the terminal strip. Many relays follow the layout shown on Figure 9-4; however, you should check this for your relay.

Set the switch into override mode (closed) and connect the power adapter. You should find that turning the variable resistor clockwise will lead to the LED coming on and the relay contacts being closed. Being in override mode, the variable resistor will be setting the temperature, so a high temperature setting that is warmer than the temperature sensed by the sensor should have the effect of turning on the heat.

You might like to experiment by adjusting the variable resistor until the LED is just on, and then putting your finger onto the temperature sensor to warm it up. When the temperature exceeds the set point, the LED should turn off again.

That is as far as we can go with testing until we have modified the home automation controller.

Step 7. Modify the Home Automation Controller

We need to add the RF transmitter to the perfboard for the home automation controller. Figure 9-11

View from the bottom of the board

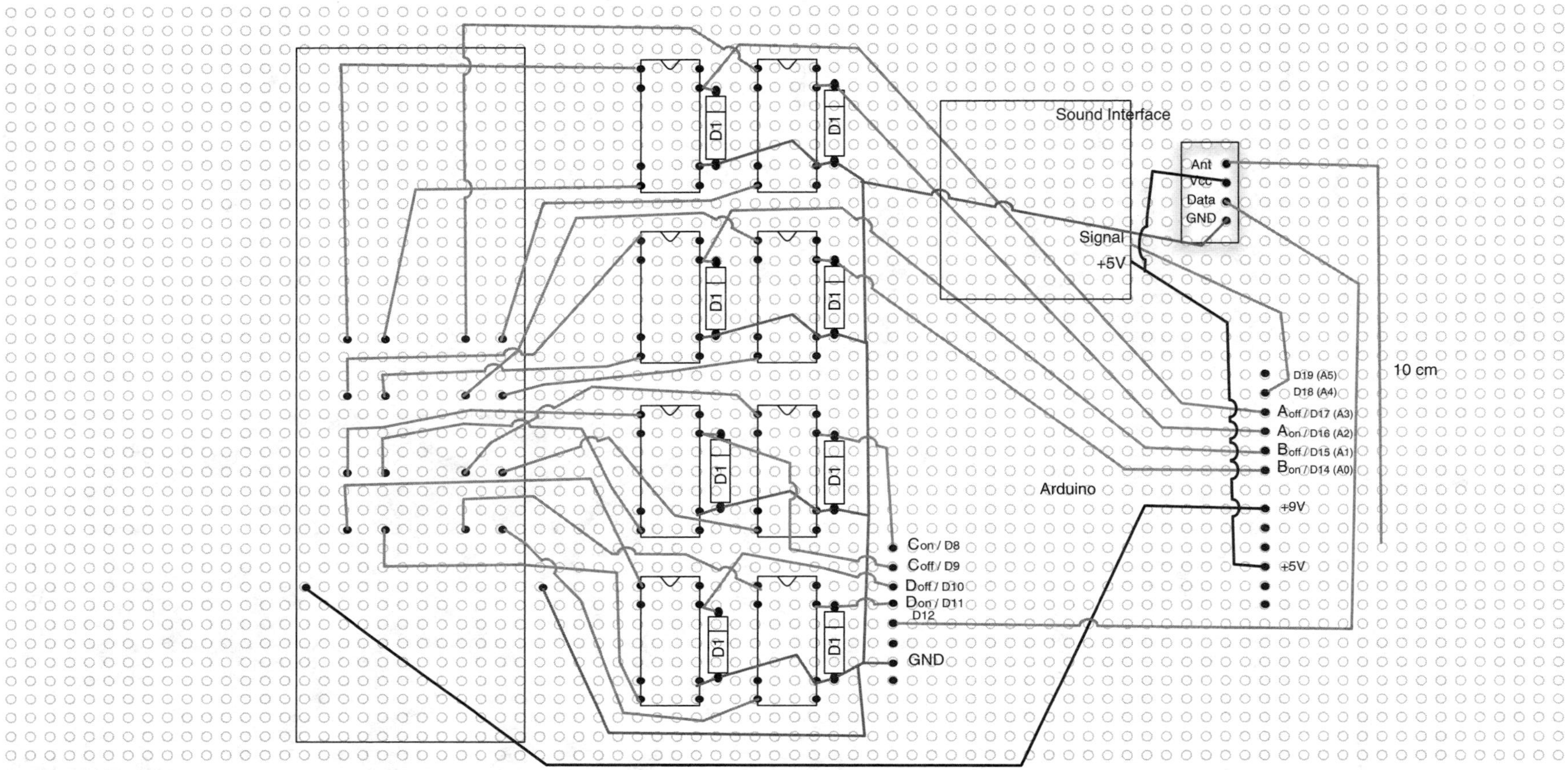

Figure 9-11 The modified perfboard layout

shows the modified home automation controller perfboard layout.

The transmitter module is smaller than the receiver module and sits to the side of the sound interface board. The data connection from the transmitter is connected to D12 of the Arduino, with the other connections being GND and +5V to Vcc. As with the receiver, you can also connect an antenna in the form of a length of wire attached to the ANT connection of the transmitter, although this is not usually necessary over short distances.

The front and back of the modified perfboard are shown in Figures 9-12 and 9-13, respectively.

The Arduino sketch that we originally uploaded onto the board included the code for the transmitter, so we do not need to update it.

Step 8. Test

We are now in a position to test the project with the home automation controller. So, turn everything on including the home automation controller and set the switch on the thermostat so

Figure 9-12 The front of the modified perfboard

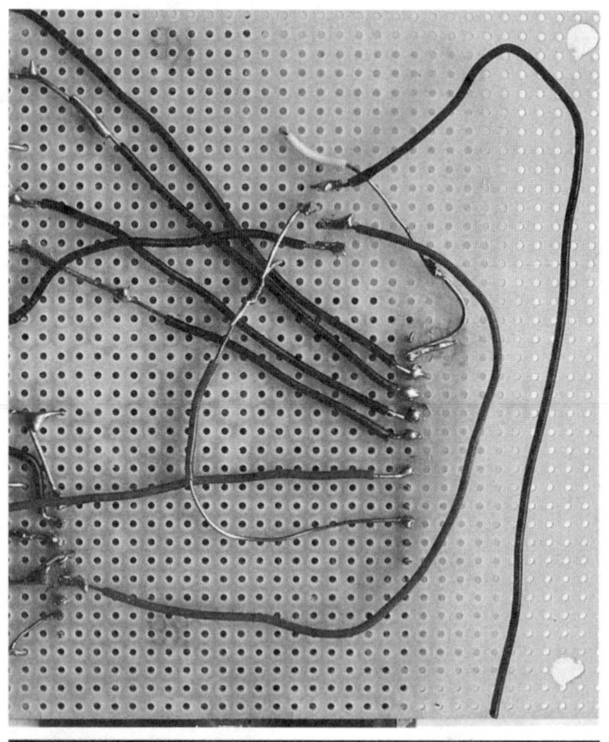

Figure 9-13 The back of the modified perfboard

that override is off. After perhaps 10 seconds, the LED should flash twice. This will indicate that the home automation controller has sent a set temperature to the thermostat.

Unless it is extremely cold, the LED and relay will be off, as the default temperature is 30° F.

Go to the appropriate "Mon–Fri Heating" page on the controller, and for the time-slot for the current time (Figure 9-17), set the temperature to the maximum temperature. After a while, the LED should flash on the thermostat and then the LED and relay switch to the on position.

If the LED does not flash, then recheck the wiring changes you made to the home automation controller and the thermostat, especially in the area of the RF modules. Check that they have 5V across their power pins.

You can check that the relay is working by setting your multimeter to continuity mode and attaching the leads to the screw terminals. When the relay clicks and the LED goes on, then the buzzer should sound on your meter.

Step 9. Box the Project

This thermostat is going to replace the incumbent thermostat. So a box that is the same size or slightly larger, but that still accommodates the stripboard and other components is ideal.

The power supply presents a number of options. If the old thermostat has both neutral and hot wires, then you could wire the power supply within the box. Otherwise, you will need a power outlet near the thermostat.

In boxing the project, the author's approach is to buy a box that is easily big enough to contain everything and then place the project components into the box and rearrange things until a good layout has been achieved.

Once I am sure where everything is going, I mark the box for drilling.

Figure 9-14 shows the components sitting in the box, while Figure 9-15 displays the drilled front panel.

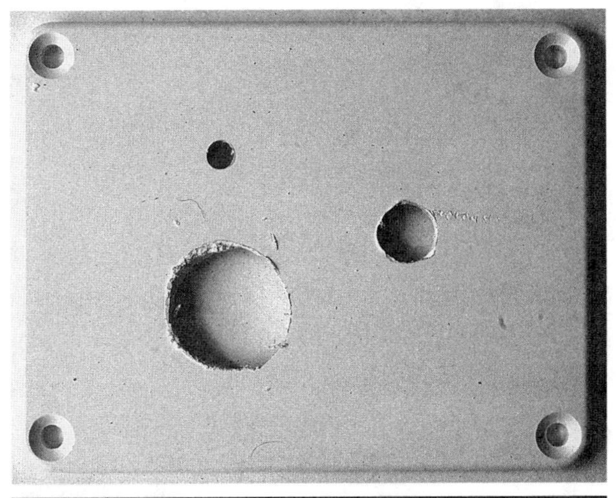

Figure 9-15 The drilled front panel

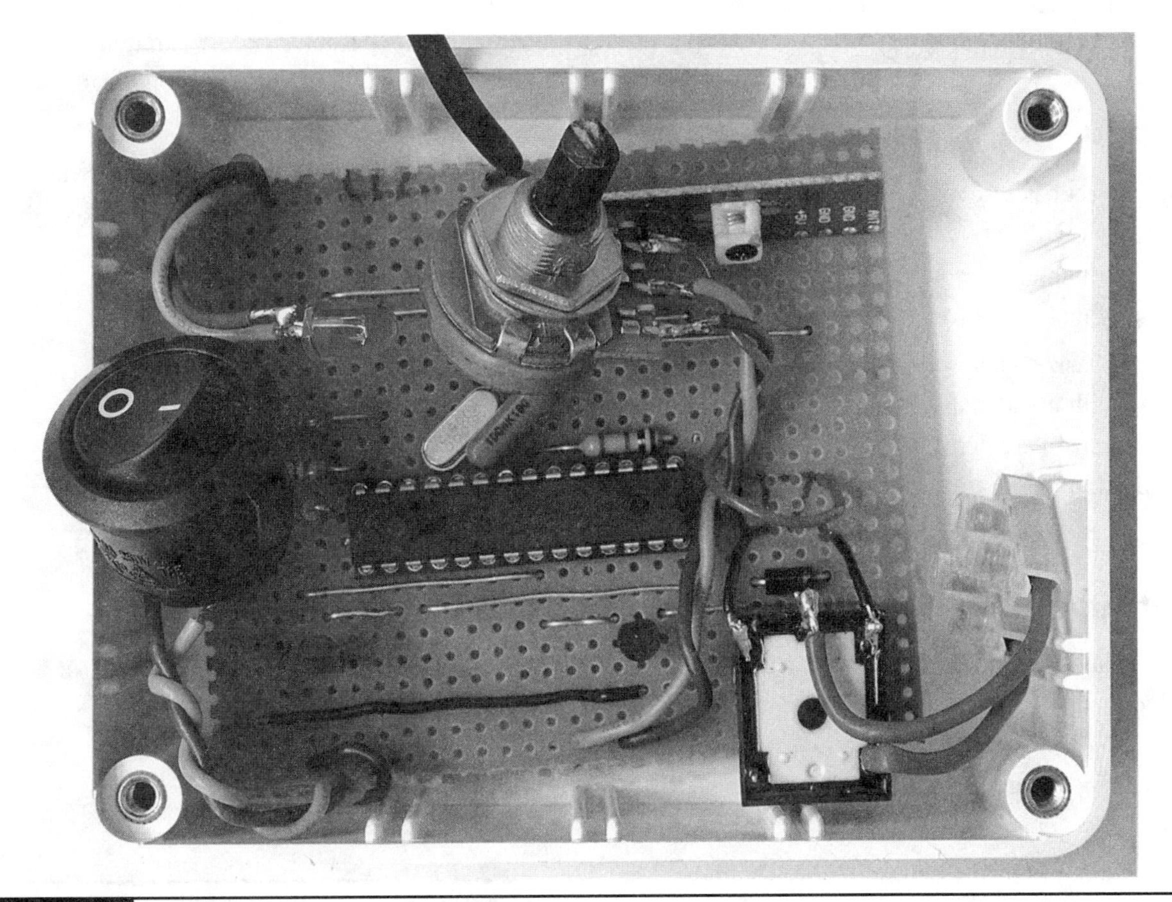

Figure 9-14 Positioning the controls

The LED is pushed through a 5mm hole. The tight fit will keep the LED in place.

You will also need to drill a hole for the lead from the power adapter, as well as some holes in the back of the box to allow the wiring from the heating system to enter and to fix the box to the wall.

Fixing holes should ideally reuse the holes in the wall used by the old thermostat.

You also need to add some temperature markings to the thermostat. If you are using degrees Fahrenheit, then the minimum temperature is 30 and the maximum 100. Mark these positions and then interpolate the six points in between (40, 50, 60, 70, 80 and 90). You can just write the numbers on the plastic case using a marker pen, or use run-on transfers, or print an underlay onto paper, cut it out, and stick it onto the box. Suitable underlay designs for both degrees Celsius and Fahrenheit can be found at this book's web site at www.automatedlair.com.

Step 10. Installation

This project has essentially provided you with a switch that will close when heating is required. This is the normally open contacts of the relay. The actual wiring of this to your central heating is up to you.

I emphasize again that you should carry out such wiring with great caution, and if you are not exactly sure what you are doing, then find someone who is qualified to work on your heating system.

Using the System

The settings page on the home automation controller includes an option to set the temperature units (Figure 9-16). If you check the box "Use degrees C," all the temperatures displayed on the home automation controller will be in degrees Celsius.

You should set this setting to your desired option before you start making your temperature profile, because if you change it afterward, you will have to do it again.

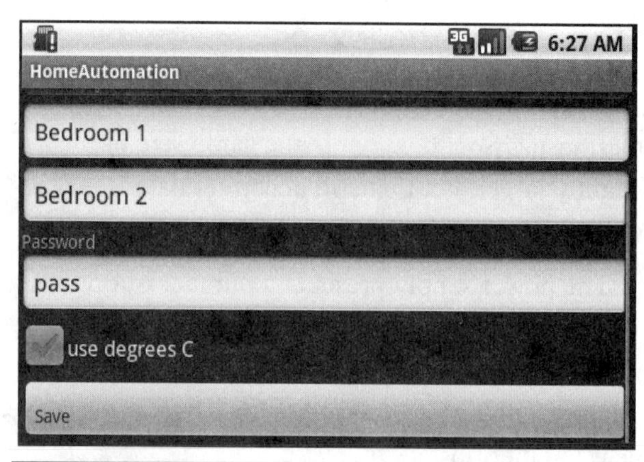

Figure 9-16 Setting the units of temperature

The Evil Genius can reduce their energy costs by setting a temperature profile that matches their daily routine. So there is little point in heating the lair if the minions and the Evil Genius himself are all out performing nefarious deeds. Since the Evil Genius prefers being a homebody on the weekends, a different schedule is required, hence the separate options for Monday to Friday compared to weekends.

Figure 9-17 shows a typical weekday temperature profile.

The temperature is set to the minimum (30° F) overnight, but then between 4 a.m. and 6 a.m. it is set to increase to 50 °F, and then to 60 °F between 6 a.m. and 7 a.m. It then goes up to 70° F between

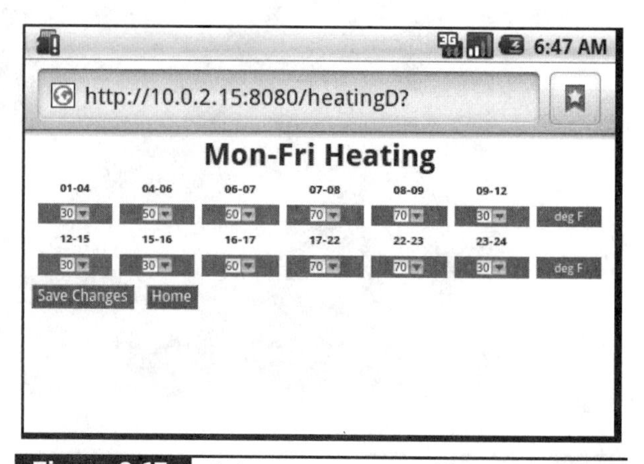

Figure 9-17 A temperature profile

7 a.m. and 9 a.m. when the Evil Genius is performing his ablutions.

From 9 a.m. until 4 p.m., when everyone is out, the temperature is reset to the minimum, rising again for the rest of the evening until it is time for bed.

Theory

This is a fairly standard Arduino type project. We do, however, use a couple of interesting components in the form of the temperature sensor chip and the RF module for Arduino-to-Arduino communication.

We will now discuss both these components, and then look at how to program an Arduino to use them.

One-Wire Sensors

The temperature sensor used in this project is a very interesting device. It combines a temperature sensor and a processor that sends the temperature in digital form. This allows it to be positioned a distance away from the Arduino, without the length of the wire causing problems with the accuracy of the temperature measurement.

See the article at www.arduino.cc/playground/Learning/OneWire for an interesting introduction to these sensors.

The Thermostat Sketch

The sketch for the thermostat is listed next.

```
#define THERMOSTAT_ID 0x41
#define MIN_ON_TIME 120L // seconds
#define RADIO_CHECK_PERIOD 10000L
#define MIN_TEMP 30
#define MAX_TEMP 100

#define rfRxPin 11
#define tempRxPin 10
#define potPin 0
#define ledPin 8
#define relayPin 7
#define overrideSwitchPin 9

OneWire oneWire(tempRxPin);
DallasTemperature sensors(&oneWire);
DeviceAddress thermometer;

int setTemp = 0;
int actualTemp = 0;
long lastCheckedRadio = 0;

void setup(void)
{
  pinMode(ledPin, OUTPUT);
  pinMode(relayPin, OUTPUT);
  pinMode(overrideSwitchPin, INPUT);
  digitalWrite(overrideSwitchPin, HIGH); //turn on pullup R
  Serial.begin(9600);
  vw_set_ptt_pin(5); // out of the way
  vw_setup(2000);
  vw_rx_start();
```

(continued)

```
  sensors.getAddress(thermometer, 0);
  sensors.begin();
  sensors.setResolution(thermometer, 10);
  Serial.println("Ready");
}

void loop()
{
  if (digitalRead(overrideSwitchPin) == LOW)
  {
    // no override, so receive temperature from Home Controller via RF
    if (millis() > (lastCheckedRadio + RADIO_CHECK_PERIOD))
    {
      checkForMessage();
      lastCheckedRadio = millis();
    }
  }
  else
  {
    setTemp = readSetTemperature();
  }
  actualTemp = readTemperature();
  setPower();
  delay(500);
}

void checkForMessage()
{
  uint8_t buf[VW_MAX_MESSAGE_LEN];
  uint8_t buflen = VW_MAX_MESSAGE_LEN;
  if (vw_get_message(buf, &buflen))
  {
    // there was something to receive.
    // We only care about the first two bytes
    // first byte identifies the receiver of the message
    // the second is the temperature in F
    byte receiver = buf[0];
    byte payload = buf[1];
    if (receiver == THERMOSTAT_ID) // A
    {
      // the message is for me
      setTemp = payload;
      Serial.print("Radio set temp to: "); Serial.println(setTemp);
      flash(2);
    }
  }
}

int readTemperature()
{
  sensors.requestTemperatures();
  float tempC = sensors.getTempC(thermometer);
```

```
    return (int)(DallasTemperature::toFahrenheit(tempC));
}

int readSetTemperature()
{
  int raw = analogRead(potPin);
  int t = map(raw, 0, 1023, MIN_TEMP, MAX_TEMP);
  return t;
}

void setPower()
{
  static boolean lastOnOff = false;
  static long powerLastChanged = 0;
  static int lastSetTemp = 0;
  boolean onOff = (actualTemp < setTemp);
  long t = millis();
  long t2 = powerLastChanged + MIN_ON_TIME * 1000L;
  boolean enoughTimeElapsed = (t > t2);
  boolean tempSettingChanged = (abs(setTemp - lastSetTemp) > 1);
  digitalWrite(ledPin, onOff);

  if ((onOff != lastOnOff) && (enoughTimeElapsed || tempSettingChanged))
  {
    Serial.print("set:      "); Serial.println(setTemp);
    Serial.print("temp:     "); Serial.println(actualTemp);
    digitalWrite(relayPin, onOff);
    powerLastChanged = t;
    lastOnOff = onOff;
    lastSetTemp = setTemp;
  }
}

void flash(int n)
{
  for (int i = 0; i <= n; i++)
  {
    digitalWrite(ledPin, HIGH);
    delay(100);
    digitalWrite(ledPin, LOW);
    delay(100);
  }
}
```

The three "include" statements load the libraries used by this sketch.

The THERMOSTAT_ID constant mirrors the code in the Android application for the home automation controller and is used to identify messages as being for the thermostat.

We then have four constants that you can change to suit your own requirements.

MIN_ON_TIME is the minimum time that the heating will be switched on or off for. This prevents "hunting," where just as you get to the set temperature point the heating would be turned on and off quite frequently. This is not good for the heating. We have set this to two minutes.

The RADIO_CHECK_PERIOD is the time between checking for an incoming signal from the home automation controller. This is set to 10,000 milliseconds or 10 seconds.

MIN_TEMP and MAX_TEMP determine the temperature range (in degrees Fahrenheit). Note that even though the sketch assumes degrees Fahrenheit throughout. If you work in degrees Celsius, you do not have to change anything in the sketch, just calibrate the dial differently.

We then define the pins used by the various devices connected to the microcontroller and initialize the OneWire and DallasTemperature libraries.

The "setup" function does the usual setting up of pins. A neat trick with the "overrideSwitchPin" is to set it as an INPUT and then do a "digitalWrite" to HIGH. This has the effect of turning on an internal pullup resistor so we do not need a separate resistor. We also start the radio listening and set up the temperature sensor.

The "loop" function has two paths through it, one if the override switch is set and one if it isn't. If the override switch is NOT set, then we check for a message from the RF receiver, as long as sufficient time has elapsed since we last checked. Otherwise, we read the set temperature from the variable resistor.

The main loop then reads the actual temperature and calls the function "setPower" to determine if the relay should be turned on or off.

The "checkForMessage" function looks to see if there is a message in the buffer from the radio receiver, and if there is, it checks that the first byte contains 0x41. This indicates that the message is intended for the thermostat, and that the next byte will contain the set temperature in degrees Fahrenheit. Every time this happens, the LED flashes twice.

The function "readTemperature" requests a temperature from the sensor and converts the reading in degrees Celsius to Fahrenheit.

The next function, "readSetTemperature", converts a reading from the analog pin to which the variable resistor is connected into a temperature in the range of 30 to 100 using the map function. This built-in function is very useful for converting readings from one range to another. The arguments it takes are the number to be converted, the range of unconverted values (in this case, 0 to 1023 from the analog input), and the range of output numbers (in this instance, 30 to 100).

Finally, the "setPower" function changes the state of the relay if required. This function uses three static variables that retain their value each time the function is called. These allow the function to determine if the state of the relay has changed, how long it is since it last changed, and the last set temperature. Only if the state has changed from on to off or vice versa and enough time has elapsed or the set temp has changed will the state of the relay be changed. This last clause of testing if the set temperature has changed is to allow a quicker response if the user changes the set temperature using the control knob.

Summary

The Evil Genius can now lie in bed with his Android phone, turning lights and power outlets on and off as well as controlling the temperature of his lair.

The next and final chapter on home automation in this book will allow the Evil Genius to prevent his minions from wandering off by building an electric door system, one that is controlled both by RFID tags and the home automation controller.

CHAPTER 10

RFID Door Lock

NO MATTER HOW MANY TIMES the Evil Genius punishes them, the minions still manage to misplace or forget their keys and get locked out of the lair. With this project (Figure 10-1), not only do the minions simply have to take an RFID key with them, which can be attached to their person, but if they also manage to lose the RFID key, the Evil Genius can unlock the door remotely using the home automation controller.

If you are intending to build this project into your home for real, you may wish to consider that although the basic RFID security mechanism and door latch probably offer better security than a regular lock, being able to unlock the door over the Internet or even just your network offers some weaknesses. So you may wish to disable that part of the project.

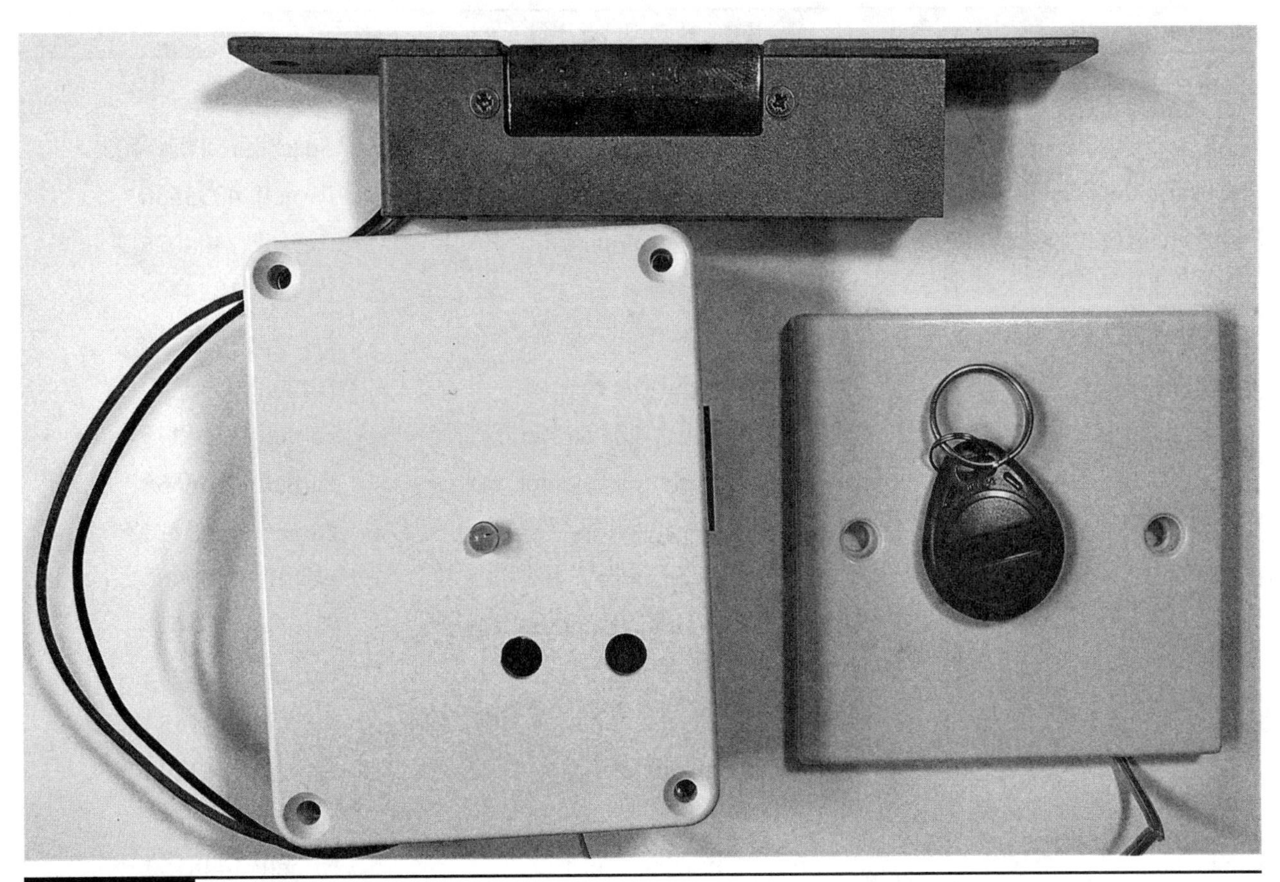

Figure 10-1 The RFID and wireless door lock

There is an element of "security through obscurity," but someone who also has this book could gain access pretty easily.

For instance, to unlock the door using the web interface, all you need to do is guess the password, or if they have an Arduino and an AM transmitter, transmit the default unlock code. If this hasn't been changed, they could gain access very easily. Even if it has been changed, there are only about 65,000 combinations, so they could try all of them automatically.

So these are the risks. You can make your own assessment and decision as to how best to protect all your worldly goods with this in mind.

Construction

Figure 10-2 shows the schematic diagram for the project. This is another project based on taking an Arduino project off-board. In many ways, it is a project quite similar to the thermostat. It uses the same kind of RF receiver, but instead of a relay it has a power MOSFET transistor controlling the door latch.

The door latch requires a 12V power supply, so we will use a voltage regulator IC to generate a 5V supply for the microcontroller. It also has a three-color LED. These LEDs are actually two LEDs—one red, one green—in the same package, with their cathodes (negative sides) connected together. You can turn on the red and green separately, and also turn them on together, in which case you get the third color, orange.

What You Will Need

You will need the following components listed in the Parts Bin to make the RFID key and door latch.

PARTS BIN

Part	Quantity	Description	Source
Microcontroller	1	ATMega328 with bootloader	SparkFun: DEV-10524
Electric door latch	1	12V DC door latch	Farnell: 4333550
RF receiver	1	433-MHz receiver module	Farnell: 1304026
IC socket	1	28-pin DIP IC socket	Farnell: 1824463
R1, R2	2	1kΩ 0.5W metal file resistor	Farnell: 9339779
R3, R4	2	270Ω 0.5W metal file resistor	Farnell: 9340300
C1	1	100μF 16V electrolytic capacitor	Farnell: 1136275
C2	1	1μF 16V electrolytic capacitor	Farnell: 1236655
C3	1	100nF ceramic capacitor	Farnell: 1200414
C4, C5	2	22pF ceramic capacitor	Farnell: 1600966
D1	1	Three-color common cathode LED	Farnell: 1581197
D2	1	1N4001	Farnell: 1458986
T1	1	FQP33N10 N-channel power MOSFET	Farnell: 9845534
IC1	1	7805 voltage regulator	Farnell: 1261233
X1	1	16-MHz crystal	Farnell: 1611761
S1, S2	2	PCB push-button switches	Farnell: 1448152

(continued)

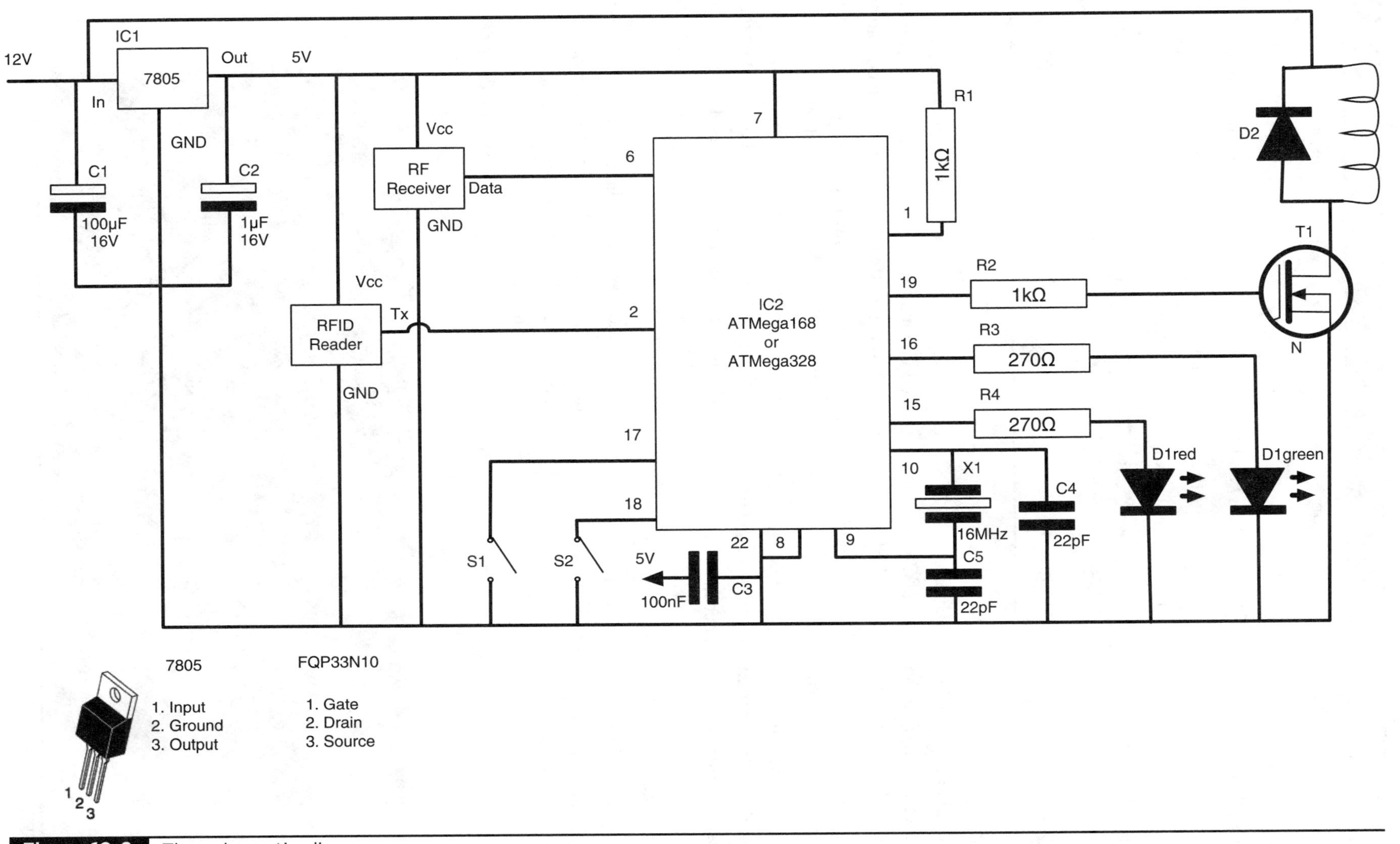

Figure 10-2 The schematic diagram

PARTS BIN *(continued)*

Part	Quantity	Description	Source
Stripboard	1	29 strips of 24 holes	Farnell: 1201473
Header socket	1	For the RFID and RF modules	Farnell: 1218869
Header plug	1	Two-way for the RFID antenna	Farnell: 1097954
Power supply	1	12V 1A power supply	Farnell: 1279478
Blanking plate	1	To enclose the RFID antenna	
Project box	1		Local electronics store
Power socket	1	2.1mm power socket	Farnell: 1217038

You will also need an Arduino Uno or Duemilanove to program the ATMega328 or ATMega168 chip. When you buy the ATMega, make sure you get a version with the Arduino bootloader already installed on it.

This project uses the same 433-MHz RF module we used in the thermostat project. In this case, you do not need another transmitter, because the same transmitter on the home automation controller can send to both the thermostat and the door lock.

The door solenoid requires a 12V DC supply at a few hundred milliamps, so we will use a 12V power adaptor with an output current of up to 1 amp.

The RFID module was bought on eBay (the Evil Genius loves a bargain) as a kit that included a 125-kHz RFID module, a suitable coil antenna, and a selection of tags. The pin connections and format of the code output are shown in Figure 10-3. If you cannot find the exact same module, use a socket on the stripboard and connect a header and lead to whatever module you end up with. You may also have to adapt the sketch if the output code is different.

The door latch is of the type that fits into the door frame. When 12V is applied across its leads, it powers a solenoid that allows the side of the socket to move. It is a fail-locked arrangement, meaning that if the power fails, the door remains

Figure 10-3 The RFID module used

locked unless you use your regular key to unlock it instead of the RFID tag. This design will work with pretty much any 12V door lock mechanism, so if you think you may need something different to the lock specified in the parts list, do a little research and see what you can find.

In addition to these components, you will also need the following tools in the Toolbox.

> **TOOLBOX**
>
> - Soldering equipment
> - A craft knife and ruler
> - Multi-core wire in various colors
> - An electric drill and assorted bits
> - A multimeter
> - Woodworking tools to fit the door latch

Step 1. Prepare the Stripboard

Figure 10-4 shows the stripboard layout for the door lock.

Cut the board to the right size. The neatest way to do this is to score the board heavily with a craft knife and then break it over the edge of a table. You can also use a strong pair of scissors, but the result will not be as neat.

The prepared stripboard is shown in Figure 10-5.

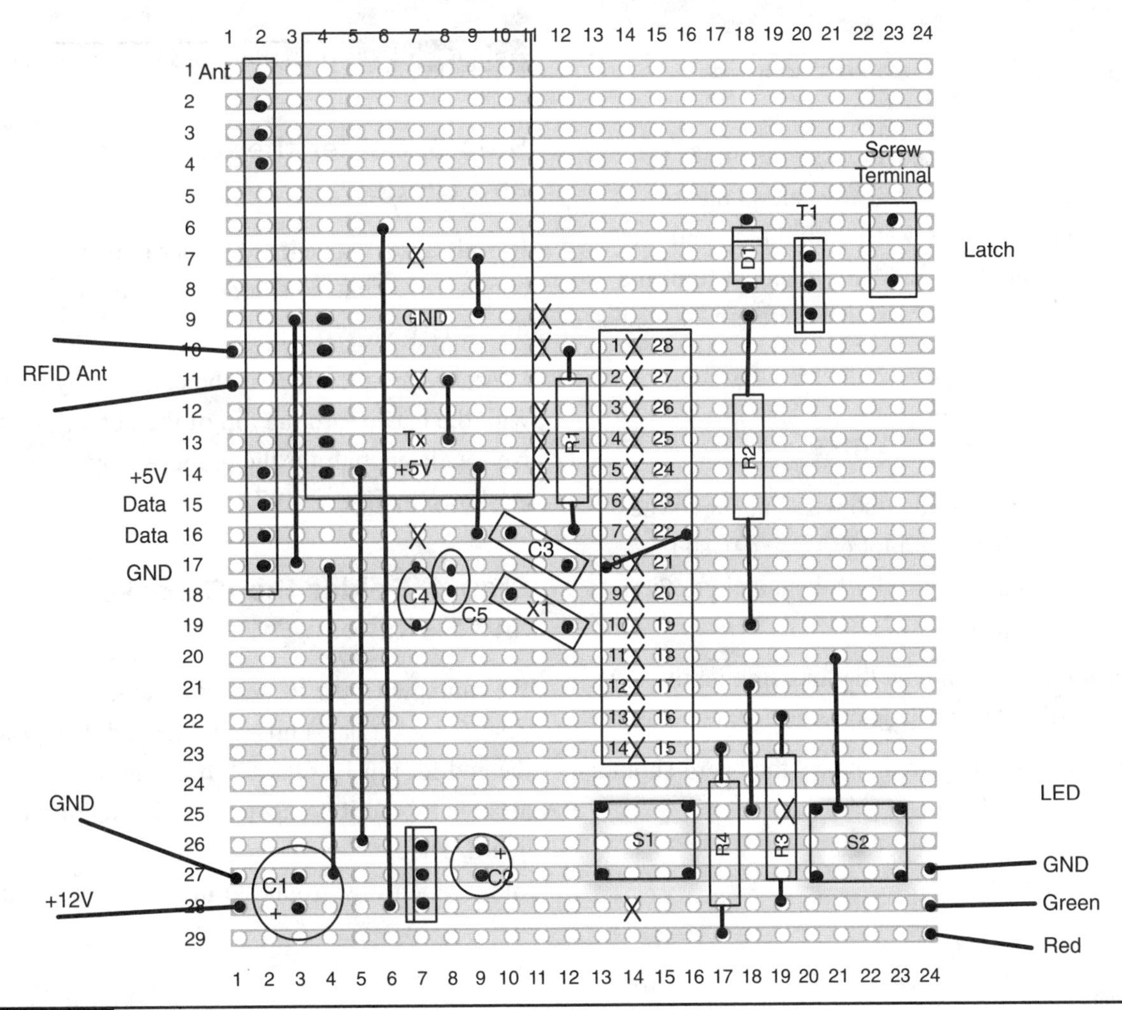

Figure 10-4 The stripboard layout

Figure 10-5 The prepared stripboard

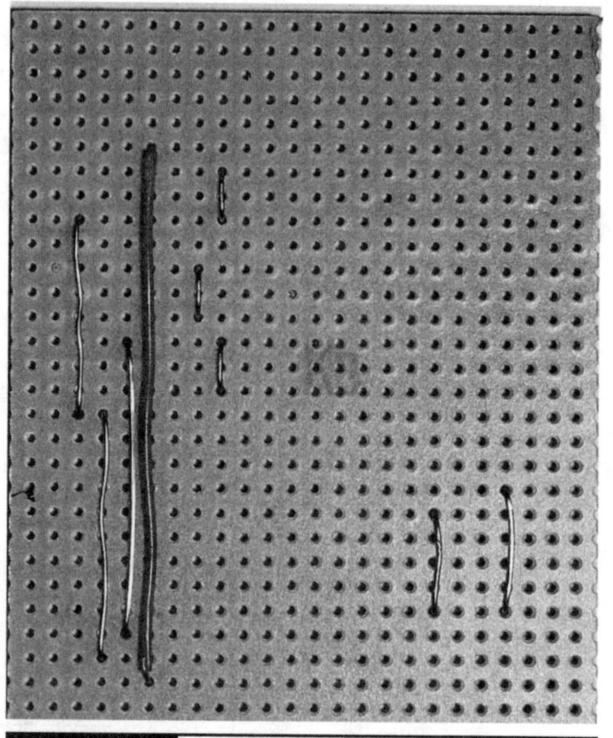

Figure 10-6 The stripboard with links in place

Use a drill bit to make breaks in the track where indicated by an "X." You can do this by just twisting the bit between your thumb and forefinger. Note that the board is shown from the top side in Figure 10-4 and that there are 24 breaks to make.

Step 2. Solder the Links

The next step is to solder the wire links onto the board. Again, there are a lot of these (9), so take care that you have made all of them and got them all in the right places.

Leave the insulation on the long link on the left of the board to protect it against accidental shorts.

When all the links are in place, the board should look like Figure 10-6.

Step 3. Solder the Resistors and Diode

It is always a lot easier to solder the lower-lying components first, so solder in the resistors and diode next.

Make sure the diode is the correct way around with the bar toward the top of the board. The span isn't wide enough for the diode, so lay it on its side as shown in Figure 10-7.

Step 4. Solder the IC Socket and Switches

We can now fit the IC socket and the two switches. The IC socket has a notch at one end that indicates the end for pin 1 of the IC. It is a good idea to arrange this so it points toward the top of the board. Then, when you come to insert the IC, you know which way around to insert it (Figure 10-8).

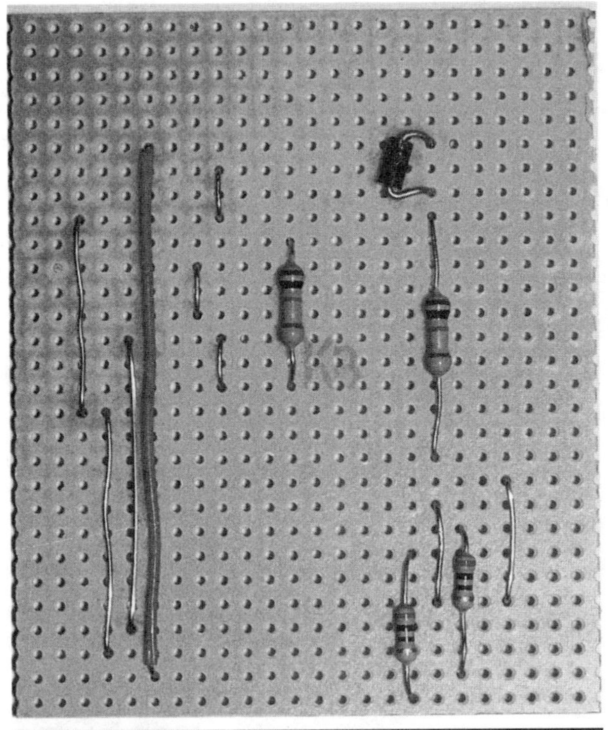

Figure 10-7 The stripboard with resistors and diode

Make sure you get the switches the correct way around. They should be oriented so that the contacts that are made when the button is pressed are at the top and bottom. You can use a multimeter set to continuity mode to check this.

Step 5. Solder the Remaining Components

The rest of the components and sockets can now be attached to the board (Figure 10-9). Note that we have not yet fitted the IC, RF, and RFID modules into their sockets.

The small capacitors can be attached either way around, but the larger can-like electrolytics must be put in the correct way around. The positive leads are longer and the negative leads are normally marked with a diamond symbol.

Similarly, make sure the transistor and voltage regulator are inserted the correct way and that you do not mix them up, given that they are in the same kind of package.

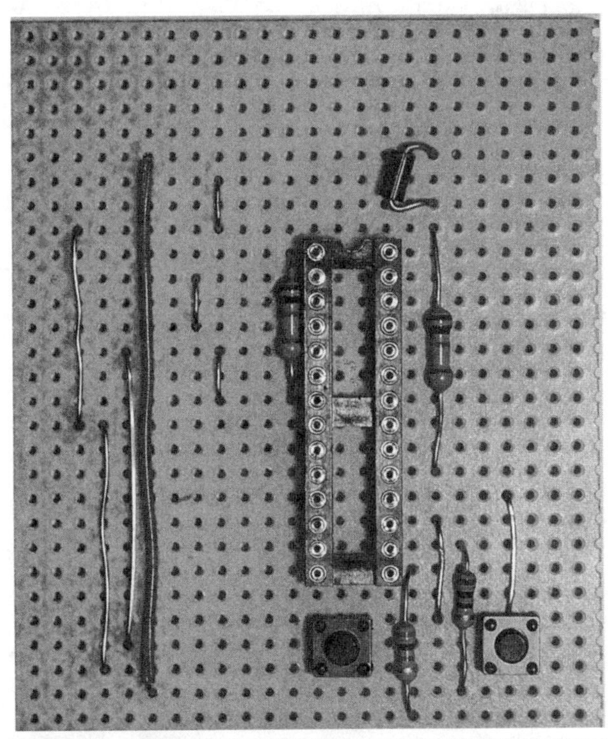

Figure 10-8 The stripboard with IC socket and switches

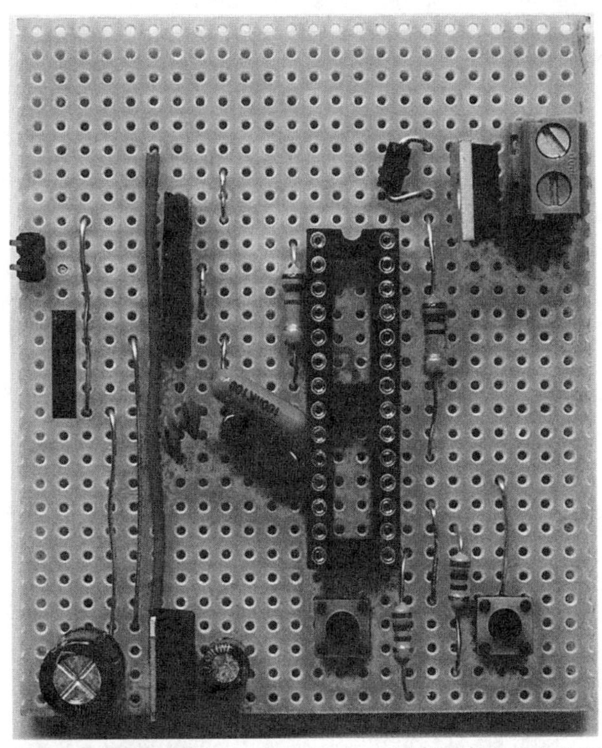

Figure 10-9 The stripboard with all the components

The author used sockets for the RF and temperature modules, but if you prefer, you can solder them directly to the board. Cut the header socket strip to make two sockets, one of four sockets for the RF module (you only need to connect the bottom half) and one of six for the RFID module. You can cut a second strip of four for the top of the RF module, but this is not essential.

A pair of header pins is used to connect the RFID antenna.

Step 6. Program and Install the Microcontroller

We need to program the microcontroller with the Arduino sketch to control the door lock. The easiest way to do that is to use an Arduino board.

It usually makes sense to program the microcontroller IC that's already in the Arduino board and replace it with the chip you bought, after the chip has been transferred to the stripboard.

Before installing the sketch, change the two constants—CODE_1 and CODE_2—to something unique for your address. Remember, this is effectively your door key. CODE_1 should be between 0x42 and 0xFF. CODE_2 can be anything between 0x00 and 0xFF. This code must match the four-digit code set in your preferences page on the home automation Android app. So if you set CODE_1 to be 0xE5 and CODE_2 to be 0x77, then you should set the code in preferences to be E577.

This project requires the use of the VirtualWire library. If you have not already installed this in Chapter 9, please follow the instructions in Step 5 of Chapter 9.

Install the modified sketch (rfid_door) onto the Arduino board. Then, with the power disconnected, carefully remove the IC from the board and fit it into the socket on the stripboard, making sure it's the right way around.

When removing the IC, be very careful as it is easy to bend or break off the pins. Using a knife, ease the IC out at each end of the socket a bit at a time, until it comes loose. You should also observe anti-static precautions, at least touching grounded metal to discharge yourself before touching the device's pins. An earthing strap is another option.

At this point, a thorough inspection of the board is a good idea. Inspect the top of the board to make sure all the components and links are in the places indicated by Figure 10-4, and check the bottom of the board to ensure all the breaks are in the right place and that there are no accidental bridges of solder between tracks.

Step 7. Connect Everything Up

Attach leads to the stripboard for the 12V supply. These can be soldered to the underside of the stripboard, or pushed through the holes from the top of the board. Use whichever method is easier for you. Attach the other end of these wires to a 2.1mm power connector and connect the three-color LED using short lengths of wire (see Figure 10-10). Check the data sheet for your LED, the one the author used had the longest lead being the common cathode that goes to ground. The shortest lead was the green anode, and the middle lead the red anode.

Slot the RFID and RF modules in place, taking care to get them the right way around.

Step 8. Test

Before we start putting things into a box, we need to test what we have made and make sure everything is as it should be. It is a lot easier to check and make any corrections before everything is boxed up.

Start by attaching the RFID antenna, the solenoid, and the power connector (Figure 10-11).

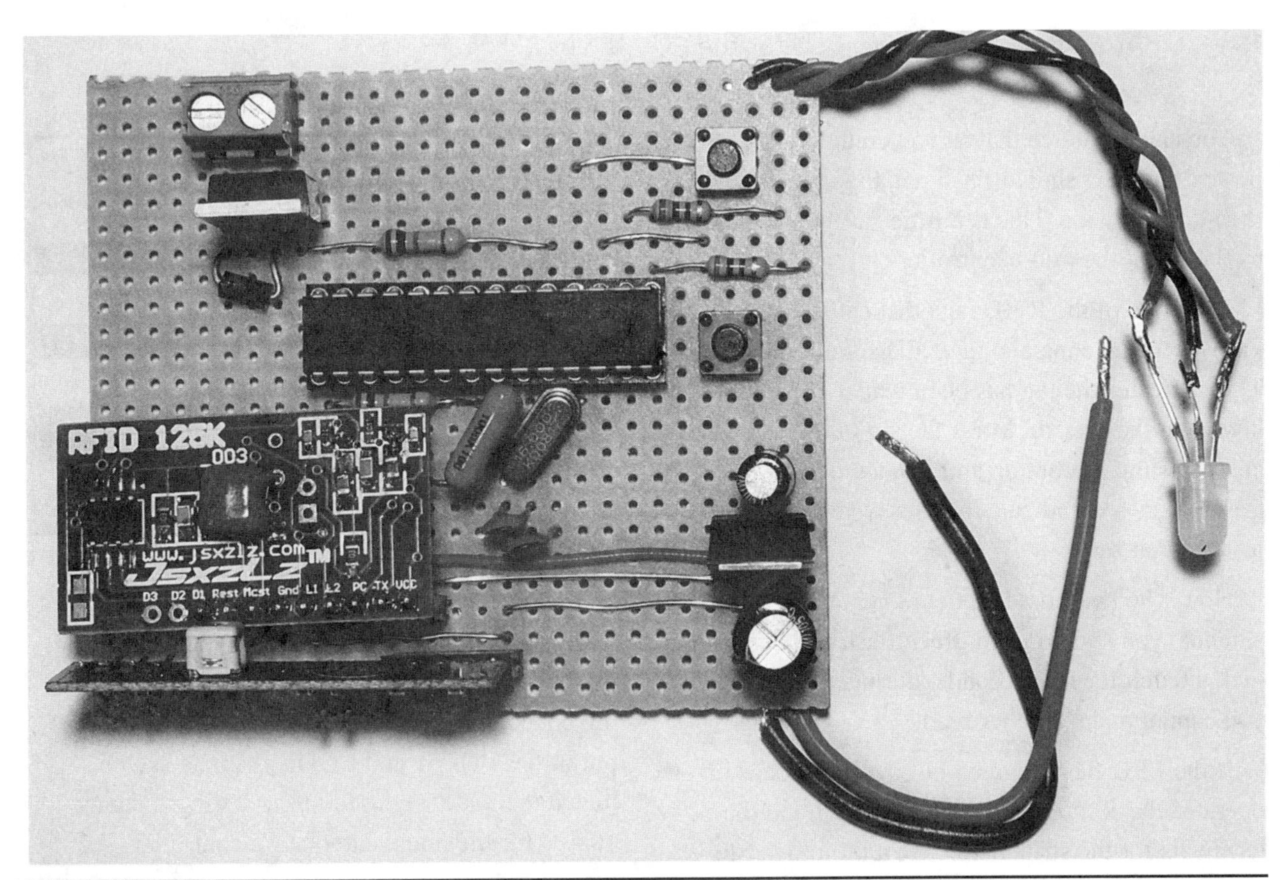

Figure 10-10 Connecting up the stripboard

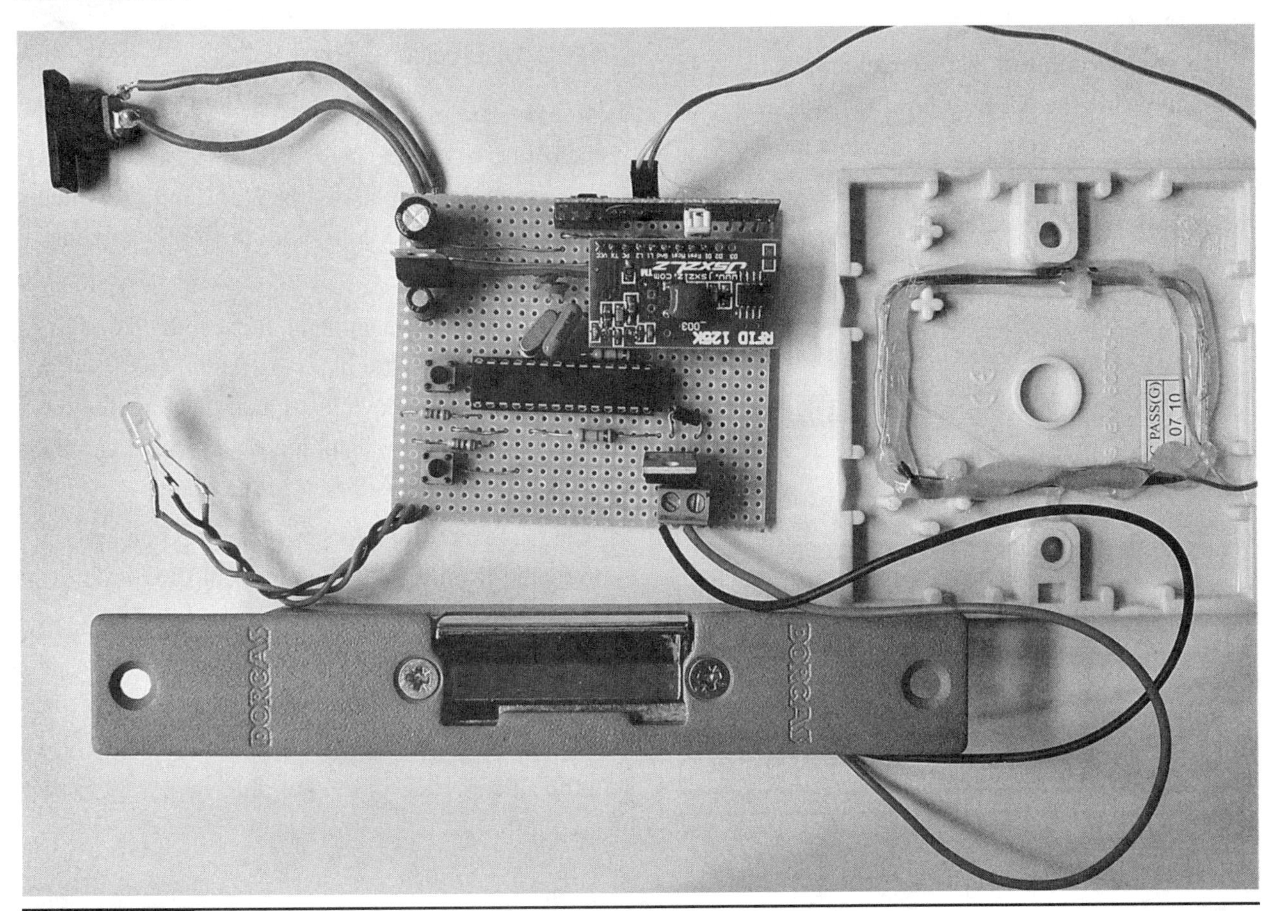

Figure 10-11 Connected up and ready to test

You should notice that as you connect the power, the LED should flash red, then green, and then orange twice. This is a little startup check just to show that the microcontroller chip is running.

Hold one of the RFID tags that came with the kit over the antenna and the LED should flash red. This indicates the tag has been read, but isn't known to the system. Move the tag out of the way and press the left button and you should get four green flashes to indicate that the system has learned that card.

Now when you hold that card next to the antenna, you should get a green flash and hear the lock unlatch for five seconds, during which time you cannot make another scan.

If the LED flashes green but the lock does not operate, check the stripboard around T1 and the terminals for the stripboard. Use a multimeter to check that there is 12V at the top connection to the lock.

We also need to test that we can unlock the door remotely using the RF link. To do this, fire up your home automation controller and select the Door option from the main menu (Figure 10-12), enter the password (the default is "password"), and click Unlock. You should hear the latch click open for five seconds.

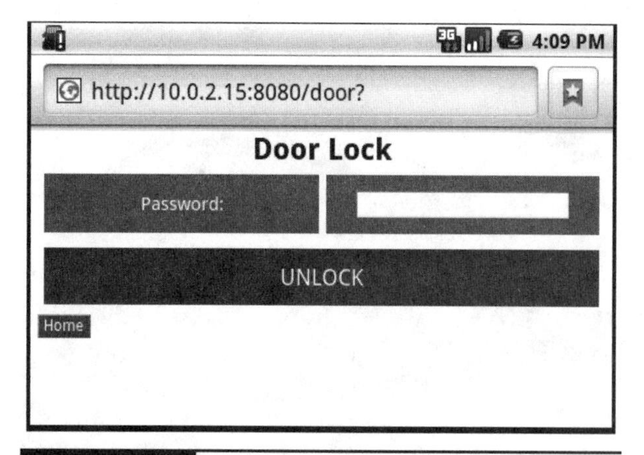

Figure 10-12 Remotely unlocking the door

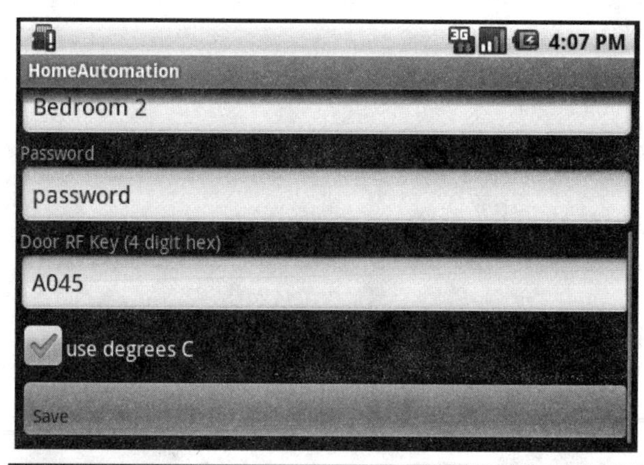

Figure 10-13 Setting the door code

If it does not, but you see an orange flash from the LED, that means it received a signal but that the code was wrong. So, make sure the code you put into CODE_1 and CODE_2 in the sketch match the code you used in the settings (Figure 10-13) for the home automation controller (see Step 6 earlier).

Step 9. Box the Project

We can now go ahead and fit everything into an enclosure, or rather enclosures, since we will need a box with the stripboard, the LED, and other items on the inside of the door, with the RFID antenna on the outside.

Starting with the antenna, this can be fitted to the inside of a plastic blanking panel, as shown in Figure 10-14. We have just glued the antenna to the underside of the blanking plate, which can then be screwed to the wall next to the door.

This is fine if the door is under a porch and protected from the weather, but if it's exposed to the elements, you will need to seal it or find a more weatherproof enclosure.

The box for the inside of the door just uses a standard project box. You will need to make holes in the side or back of the box for:

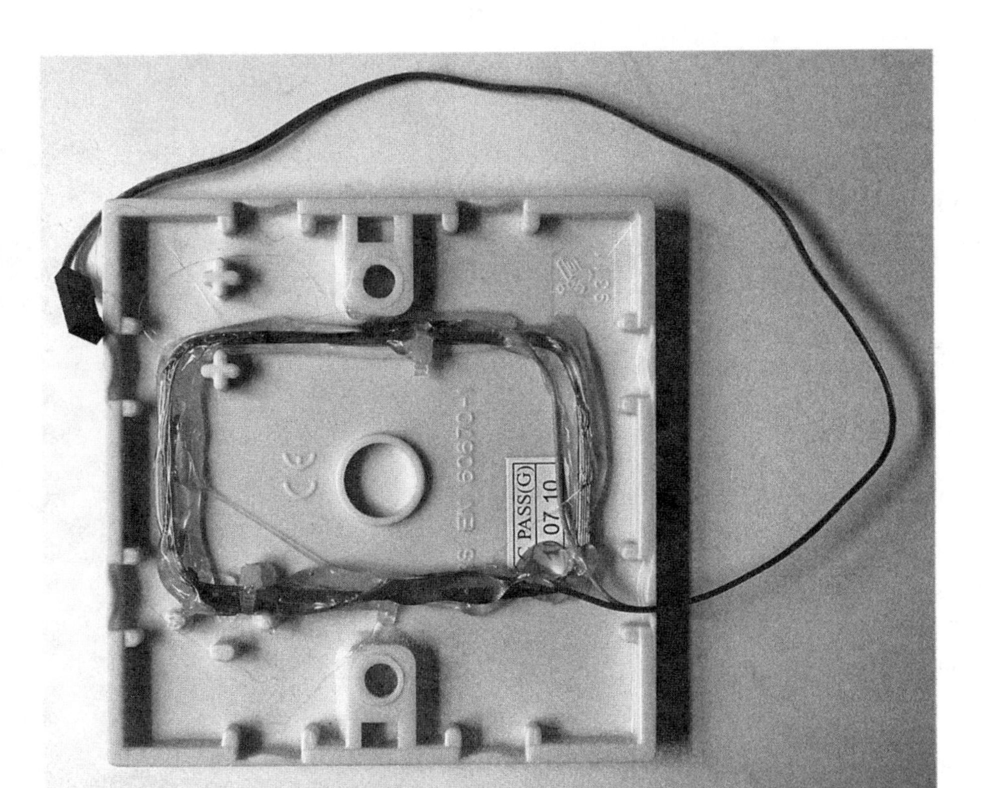

Figure 10-14 Encasing the RFID antenna

- The RFID antenna lead
- The 2.1mm power socket
- The door latch wires
- Attaching the box to the wall

The front of the box lid will have a 5mm hole for the LED and two holes immediately above the two switches into which a pen or other plastic object can be inserted to press the buttons. Figure 10-15 shows the lid and main box of the body drilled. Lay out your components in the box and mark the position of the holes you need to make before drilling.

The buttons are deliberately made a bit inaccessible to discourage the minions from programming cards for their visiting friends.

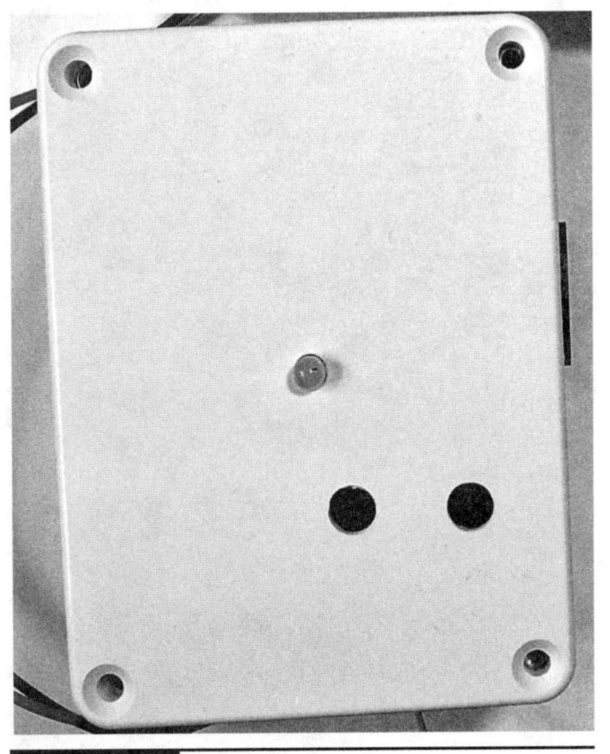

Figure 10-15 The drilled enclosure

Step 10. Installation

The procedure for fitting the latch will depend on the type of latch you have bought. If it is similar to the one recommended in the parts bin, you will almost certainly need to chisel out a bigger hole in your door frame. You will also need to drill a hole all the way through the wall or door frame for the wires to the RFID antenna. This should be directly opposite the control box on the inside of the house.

The control box should be close to the door latch itself to accommodate the leads to the electric lock.

Figure 10-16 shows the control box (without a lid), and Figure 10-17 displays the RFID antenna and the door latch itself.

Figure 10-16 The control box

Figure 10-17 The RFID antenna installed outside the door

Using the System

The system is easy to use. There are only two buttons. The left-hand Add button adds the last tag read to the list of tags that will unlock the door, and the right-hand button clears out the memory and forgets all the tags.

The three-color LED is used to good effect to provide feedback about what the system is doing. The various multicolor flashes it makes are summarized in Table 10-1.

TABLE 10-1 Summary of Multicolor Flashes

Flash	Meaning
Two orange flashes	Signal received from the home automation controller, not intended for the lock. Useful to see that the home automation controller is working.
Four orange flashes	Unlock command received from the home automation controller. This will quickly be followed by a green flash.
Green flash	The door is unlocked.
Red flash	An invalid card has been read.
Two green flashes	Following the pressing of the Add button, this indicates the tag was already known.
Four green flashes	Following the pressing of the Add button, this indicates the tag was not previously known and has been added to EEPROM.
Five red flashes	Following the pressing of the Add button, this indicates that all 16 memory slots have been used, so the tag has not been stored.
10 fast red flashes	This indicates that the Clear button has been pressed and all 16 slots cleared.

Theory

In this section, we will look at the Arduino sketch used in this project, especially how we store the RFID code in EEPROM.

The Door Lock Sketch

The sketch for the door lock is relatively long and complex in comparison to the sketches for other projects in this book. The complexity comes about in part because of the need to be able to train the Arduino on valid RFID codes and store them on the Arduino in such a way that they are not forgotten when power is lost.

```
#include <EEPROM.h>
#include <VirtualWire.h>
#define CODE_1 0xA0
#define CODE_2 0x45
#define doorOpenTime 5000 // 5 seconds
#define numCodes 16
#define codeSize 14
#define redPin 9
#define greenPin 10
#define doorReleasePin 13
#define addButtonPin 11
#define removeButtonPin 12
#define rfRxPin 4

#define RED 1
#define GREEN 2
#define ORANGE 3

char code[codeSize + 1];

char *EMPTY_CODE = "00000000000000";

void setup()
{
  Serial.begin(9600);
  pinMode(redPin, OUTPUT);
  pinMode(greenPin, OUTPUT);
  pinMode(doorReleasePin, OUTPUT);
  pinMode(addButtonPin, INPUT);
  digitalWrite(addButtonPin, HIGH);
  pinMode(removeButtonPin, INPUT);
  digitalWrite(removeButtonPin, HIGH);
  flash(RED, 2, 200);
  flash(GREEN, 2, 200);
  flash(ORANGE, 2, 200);

  vw_setup(2000);
  vw_set_ptt_pin(5); // out of the way
  vw_set_tx_pin(6); // out of the way
```

```
  vw_set_rx_pin(rfRxPin);
  vw_rx_start();
}

void loop()
{
  if (Serial.available() > codeSize)
  {
    readCard();
    checkCode();
  }

  if (!digitalRead(removeButtonPin))
  {
    clearAllCodes();
  }
  if (!digitalRead(addButtonPin))
  {
    addCode(code);
  }
  checkForMessage();
}

void readCard()
{
  int i = 0;
  char ch = Serial.read();
  while (ch != '\n' && i <= codeSize)
  {
    code[i] = ch;
    ch = Serial.read();
    i ++;
  }
  Serial.flush();
}

void checkForMessage()
{
  uint8_t buf[VW_MAX_MESSAGE_LEN];
  uint8_t buflen = VW_MAX_MESSAGE_LEN;
  if (vw_get_message(buf, &buflen))
  {
    // there was something to receive.
    flash(ORANGE, 1, 200);
    if (buf[0] == CODE_1 && buf[1] ==
    CODE_2)
    {
      flash(ORANGE, 2, 200);
      unlockDoor();
    }
  }
}

void checkCode()
{
 if (isValidCode(code))
 {
    unlockDoor();
 }
 else
 {
   flash(RED, 1, 500);
 }
}

void unlockDoor()
{
 clearLastCode();
 flash(GREEN, 1, 200);
 digitalWrite(doorReleasePin, HIGH);
 delay(doorOpenTime);
 digitalWrite(doorReleasePin, LOW);
 Serial.flush();
}

void clearLastCode()
{
   for (int i = 0; i < codeSize; i++)
   {
     code[i] = 'F';
   }
}

void flash(int color, int times, int
   duration)
{
  int red = color & 0x01;
  int green = color >> 1;
  for (int i = 0; i < times; i++)
  {
    digitalWrite(redPin, red);
    digitalWrite(greenPin, green);
    delay(duration / 2);
    digitalWrite(redPin, LOW);
    digitalWrite(greenPin, LOW);
    delay(duration / 2);
  }
}
```

```
boolean isValidCode(char *code)
{
   return (findCodePosition(code)
   < numCodes);
}

void addCode(char *code)
{
  if (isValidCode(code))
  {
    // code already stored
    flash(GREEN, 2, 500);
  }
  else
  {
    int pos = findCodePosition
    (EMPTY_CODE);
    if (pos != (codeSize + 1))
    {
      writeCode(pos, code);
      flash(GREEN, 4, 500);
    }
    else
    {
      // no room to store code
      flash(RED, 5, 500);
    }
  }
}


int findCodePosition(char *code)
{
  int pos = 0;
  while (pos < numCodes &&
   !codesEqual(code, pos))
  {
    pos ++;
  }
  return pos;
}

void writeCode(int pos, char *code)
{
   for (int i = 0; i < codeSize; i++)
   {
     EEPROM.write(pos * 16 + i,
     code[i]);
    }
}


void clearAllCodes()
{
  for (int pos = 0; pos < numCodes;
   pos++)
  {
    writeCode(pos, EMPTY_CODE);
  }
  flash(RED, 10, 50);
}



boolean codesEqual(char *code, int pos)
{
   for (int i = 0; i < codeSize; i++)
   {
     char ch = (char)EEPROM.read(pos
     * 16 + i);
     if (code[i] != ch)
     {
       return false;
     }
   }
   return true;
}
```

This is a well-structured sketch and a good illustration of how complex logic can be simplified by keeping your functions small.

Starting at the top, we first include the VirtualWire and EEPROM libraries. The VirtualWire library lets us receive communications from the Arduino in the home automation controller, and the EEPROM library lets us store the RFID codes in EEPROM memory so it is preserved even when the power is lost to the Arduino board.

The two constants CODE_1 and CODE_2 should be changed to something unique for your address. Remember, this is effectively your door key. CODE_1 should be between 0x42 and 0xFF. CODE_2 can be anything between 0x00 and 0xFF.

This code must match the four-digit code set in your preferences page on the home automation Android app. So, if you set CODE_1 to be 0xE5 and CODE_2 to be 0x77, you should set the code in preferences to be E577.

The constant "doorOpenTime" defines how long the door will remain unlatched when it has been released. The number is in milliseconds, so 5,000 is 5 seconds.

After the usual pin definitions, the "setup" function sets up the pins appropriately and then starts the VirtualWire library listening to communications over the radio.

The RFID reader is connected to the Rx pin of the Arduino, so whenever a tag is near enough for it to read, it will send the card's ID to the Arduino. The first thing we do in the "loop" function is to see if we have received any card data using Serial.available(). If some data are found, then the code is read and checked to see if it is a valid RFID code. After this, the loop function also checks to see if either of the buttons have been pressed and takes the necessary actions. Finally, the loop function checks for any messages received from the radio.

That description explains what's going on at the top level. Each of those actions is implemented as a series of layers of functions, which we will discuss in more detail now.

"readCard" will read a character at a time until the correct message length is reached, or there is nothing more to read, or until the "\n" newline character is read. It puts the card ID just read into the global variable "code," which is used throughout the sketch to refer to the last code read.

"checkForMessage" looks to see if a message has been received from the radio, and if one has, it checks that the code matches, and if it does, it unlocks the door.

"checkCode" simply checks the last code read to see if it is a code that the Arduino knows is valid. If it is, it unlocks the door.

There are no prizes for guessing what "unlockDoor" does. The command "Serial.flush()" clears out any subsequent card readings in the buffer that would otherwise cause the lock to repeatedly unlock, as every time a card is presented to the reader it is read multiple times.

"clearLastCode" clears the last code that was read. This is called once the door has been unlocked and if it was not there, the door would repeatedly unlock until an unknown tag was scanned.

The "flash" function is used to communicate the state of the system to the user. It takes three arguments. The first is the color, which can be RED, GREEN, or ORANGE. The second argument is the number of times to flash, and the final argument is the delay in milliseconds between each flash. This function is called from all over the sketch to indicate what is happening.

The function "isValidCode" checks to see if the code supplied as an argument is a known and valid RFID code. This is a thin layer over the function "findCodePosition", which reads though each of the possible 16 codes to see if the code supplied matches. If it matches, the position is returned. If no match is found, 16 is returned.

"addCode" is used when the lock has a new code added, as a result of someone pressing the Add button. If it already knows the code, it flashes green twice; otherwise, it finds the first unused slot in the 16 code positions and stores the code there, finally giving four green flashes as confirmation. If all 16 slots are in use, it gives five red flashes to indicate it was not possible to store the new code.

"writeCode" takes two arguments: the code to write and the position from 0 to 15 in which to write it. All reading and writing to the EEPROM has to take place a single byte at a time.

As the name suggests, "clearAllCodes" resets all 16 codes, making them available again to store tag IDs.

Finally, "codesEqual" compares a code with a code at the position specified.

Summary

This was the last of the projects for the home automation system. In the next chapter, we look at how you can use an Arduino with an Ethernet shield to carry out some simple web-based automation tasks.

CHAPTER 11

Signaling Flags

If, during the course of a leisurely bath, the Evil Genius requires further refreshment or other essentials, he can use any Internet device that is handy—perhaps his beloved ebook reader, for instance—to summon a minion from the kitchen quarters by raising a small servo-driven flag (Figure 11-1).

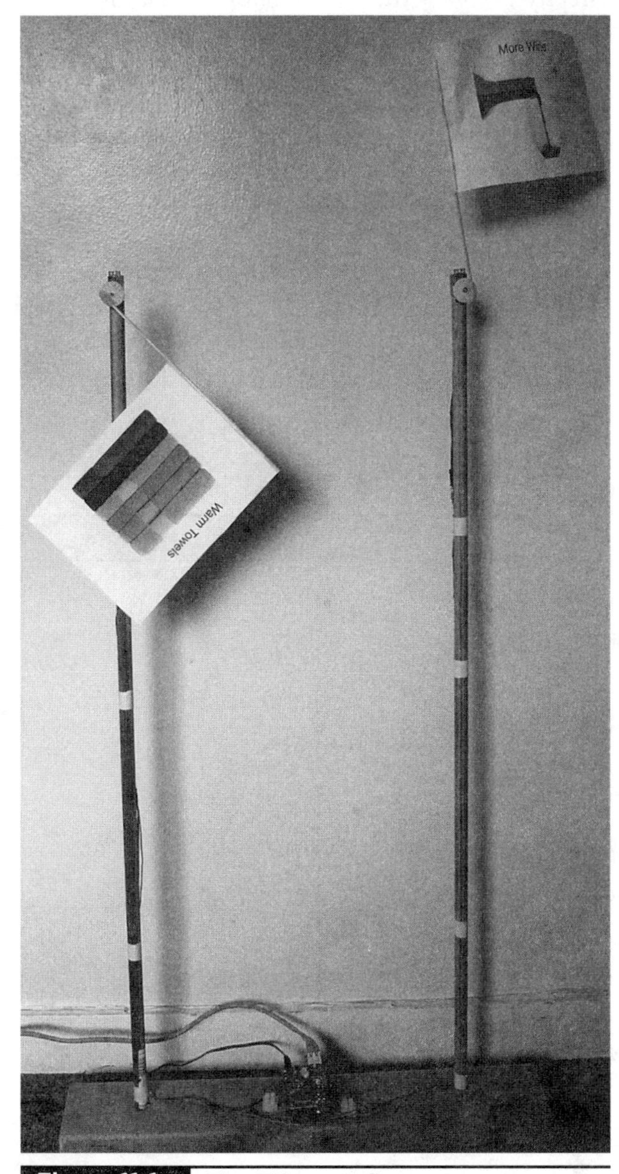

Figure 11-1 The Ethernet flags

In fact, the Evil Genius has a choice of two flags, to which different instructions can be attached. He may, for instance, choose to have one flag for requesting more wine and another for hot towels.

The flags are controlled from the web page shown in Figure 11-2.

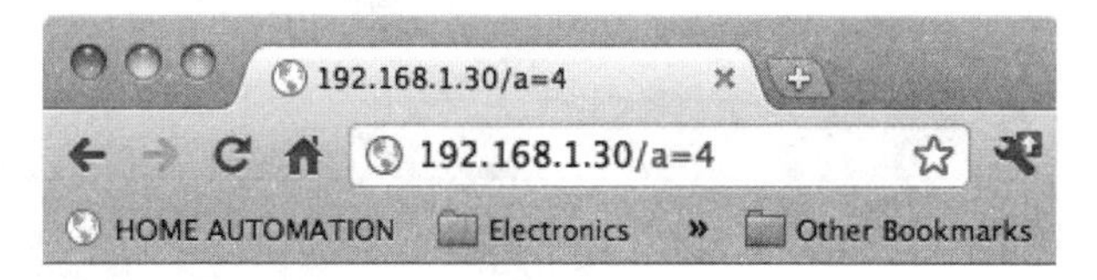

Figure 11-2 The controlling web page

This flag is strategically placed next to the butler's favorite armchair in the kitchen. Once activated, the flag will rise to the vertical position in front of the butler's nose until the servant summons up sufficient consciousness to attend to the Evil Genius.

As a secondary use, enthusiasts of the agile software development technique called eXtreme Programming have used this project to "raise a flag" while an integration is in progress, as suggested by Kent Beck.

Construction

This project is easy to build and does not require much soldering, other than attaching header pins to wires. It does, however, require a little fairly simple woodworking.

The project uses an Arduino and an Ethernet shield. This allows the Arduino to act like a tiny web server so that web requests can be sent to the Arduino from a web browser to raise or lower the flags.

The flags are mounted on small servo motors, which in turn sit on top of wooden poles. The flags are made of paper that is attached to a wooden barbecue skewer. When the servo rotates, it "raises" the flag.

What You Will Need

You will need the following components shown in the Parts Bin to construct this project.

The servo motors were obtained from eBay. Small 9g servo motors were used because they draw so little current that they can be driven directly from an Arduino output pin. Look for servo motors with a current rating of less than 40mA.

As with most of the projects in this book, you will need wire to connect things up. For this project, you will need more wire than usual because you'll have to run wires from the servos to the Ethernet shield.

In addition to these components, you will also need the following tools shown in the Toolbox.

PARTS BIN

Part	Quantity	Description	Source
Arduino	1	Arduino Uno	*See description*
Ethernet shield	1		SparkFun: DEV-09026
Servo motor	2	9g servo motor	eBay
Pin headers	1	0.1-inch header strip	Farnell: 1097954
Screw terminals	2	2A two-way screw terminals	Farnell: 1055837
Wooden base	1	2 feet (600mm) of 2 inch by 4 inch or similar	Hardware store
Wooden pole	2	3 feet (900mm) of ½-inch (12mm) dowel	Hardware store
Wooden skewer	2	10-inch (250mm) wooden barbecue skewers	
Paper	1	Letter- or A4-size paper to make the flag	
Power supply	1	12V, 1000mA or more, wall-wart type power supply	Farnell: 1279478

TOOLBOX

- An electric drill and assorted drill bits
- A wood saw
- A craft knife
- Assorted screws
- Wood glue
- A hot glue gun or epoxy glue
- A computer to program the Arduino
- A USB-type A-to-B lead

There are many possible sources for the Arduino Uno board, so shop around.

When buying an Ethernet shield, use some care because you will need an "official" shield based on the Wiznet chipset, NOT one of the cheaper but more difficult to use unofficial boards based on the ENC28J60 Ethernet controller chip.

Figure 11-3 Cutting a flat edge to mount the servo

Step 1. Construct the Wooden Platform

The size and shape of the wood really does not matter much. You just need to make sure they are far enough apart not to hit each other, and tall enough to prevent the flags from hitting the ground in the "down" position. You may find a different way of positioning the flags, perhaps making a bracket to fit on the wall.

The platform uses a good solid wooden base with about 2 feet (600mm) of thick wood, into which two holes were drilled that are the same diameter as the wooden poles. The poles were then glued into the base.

The top of the poles was trimmed on one side (Figure 11-3) with a craft knife to make a flat edge to which a server could be screwed by its bottom lug (Figure 11-4).

Figure 11-4 A servo attached to the pole

Place the Arduino board onto the center of the base and place the screw terminal blocks on each side. Mark the position of the holes for the Arduino and the terminal blocks and drill guide holes for fixing them all in place. Fix the terminal blocks in place.

Step 2. Wire the Servos to the Terminal Blocks

The schematic diagram for the project is shown in Figure 11-5, and the wiring diagram is displayed in Figure 11-6.

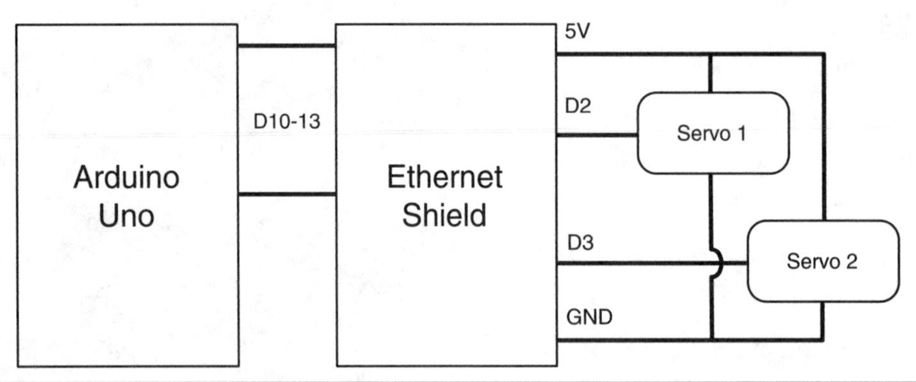

Figure 11-5 The schematic diagram for the project

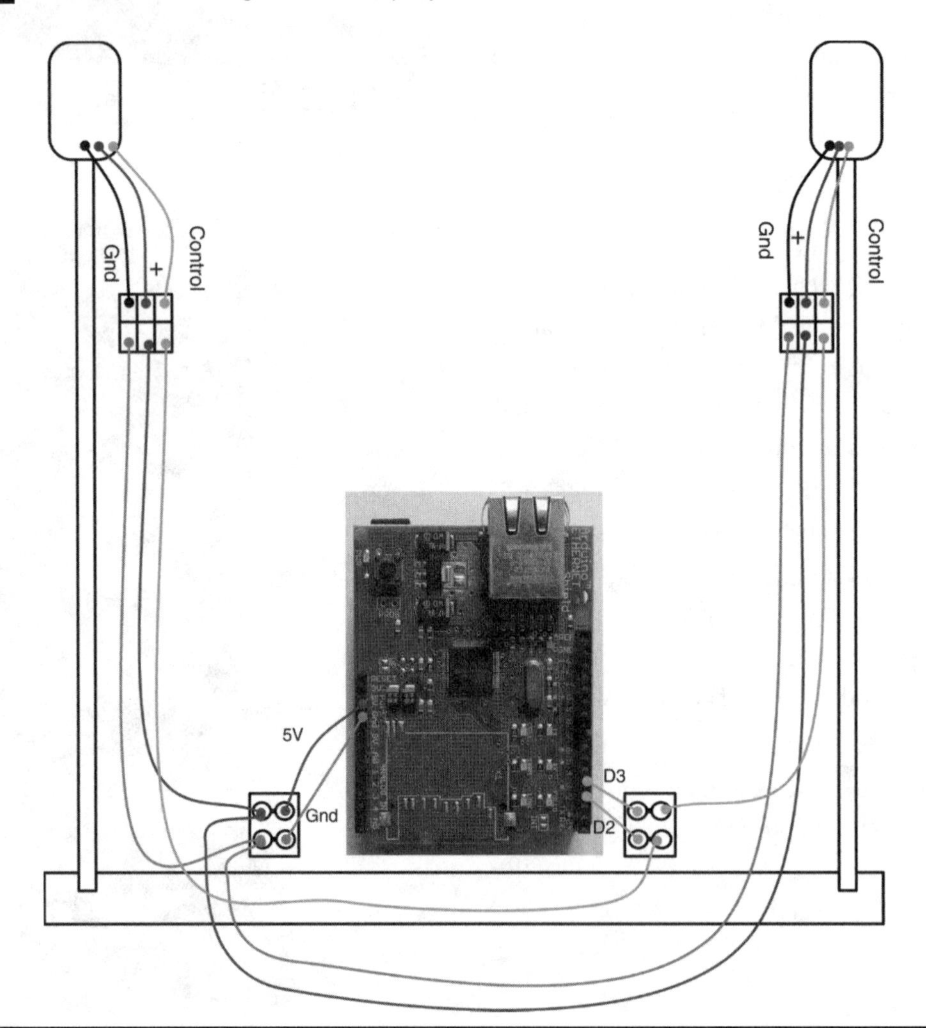

Figure 11-6 The wiring diagram for the project

The servo leads are terminated in 0.1-inch header sockets, each servo having three connections. Two connections for power (GND and 5V) and one control signal. Cut lengths of wire to span from the header sockets of the servo to the terminal blocks and solder a three-pin section of the header strip to the appropriate wires, using the wiring diagram in Figure 11-6 as a guide.

Step 3. Fix the Arduino to the Base

When the wiring between the terminal blocks and the servos is all in place, use small screws to fix the Arduino into position and place the shield on top.

Then, cut four short wires of about two inches (50 mm) and attach them in pairs to two sections of two-pin headers.

Plug the power lead header into the +5V and GND sockets on the shield and the control signal header into pins D2 and D3 on the other side of the shield (see Figure 11-6).

Step 4. Program the Arduino

Detach the shield from the Arduino before programming it for the first time. This will prevent potential damage from the pin modes set in the last sketch to be used on the Arduino board. Subsequent programmings can be carried out with the board and shield in place.

Open the sketch ch11_ethernet_flag. All the sketches can be downloaded as a single zip file from www.duinodroid.com.

Before uploading the sketch, there are a couple of changes we need to make. If you look at the top of the sketch, you will see the lines:

```
byte mac[] = { 0xDE, 0xAD, 0xBE, 0xEF,
    0xFE, 0xED };
byte ip[] = { 192, 168, 1, 30 };
```

The first of these, the mac address, just has to be unique among all the devices connected to your network. The second one is the IP address. Whereas most devices that you connect to your home network will have IP addresses assigned to them automatically by a process called DHCP, this is not true of the Ethernet shield. For this device, you have to manually define an IP address. This cannot be any old four numbers, they must be numbers that qualify as being internal IP addresses and that fit in the range of IP addresses expected by your home router. Typically, the first three numbers will be something like 10.0.1.x or 192.168.1.x, where x is some number between 0 and 255. Some of these IP addresses will be in use by other devices on your network. To find an unused but valid IP address, connect to the administration page for your home router and look for an option that says DHCP. You should find a list somewhere online of devices and their IP addresses similar to that in Figure 11-7. Select a final number to use in your IP address. In this case, 192.168.1.30 looked like a good bet and indeed it worked fine.

Attach the Arduino to your computer by USB lead and upload the sketch.

Step 5. Test

At last, we are ready to test, so put the shield in place, attach the wires from the terminal block and plug in the power supply. The servos should snap into position.

Open a connection on your browser to the IP address that you assigned for the Ethernet shield and try pressing the four buttons on the web page. You should hear the servos whir.

Figure 11-7 Finding an unused IP address

Step 6. Make and Attach the Flags

The flags are just bits of paper attached to the wooden skewers. The skewers are glued to one of the "arms" supplied with the servo. The "arms" have a cogged socket that allows the arm to be attached in any position.

To make sure the flag is in the correct position. Press the "Warm Towels" and "More Wine" buttons to put both flags up, and then fit the flags with the skewers as close to vertical as possible. Try the flags out a bit and then attach the retaining screw to each arm when you are sure that everything is working as it should.

Theory

In this "Theory" section, we will have a brief look at servo motors.

Servos

Servo motors are most commonly used in radio-controlled model vehicles.

Unlike standard motors, a servo module does not rotate round and round, but can only travel through about 180 degrees. They are controlled by pulses of voltage on the control connection to the servo. The length of the pulse controls the angle to which the servo is set.

Figure 11-8 shows an example waveform. You can see how the width of the pulses varies the servo angle.

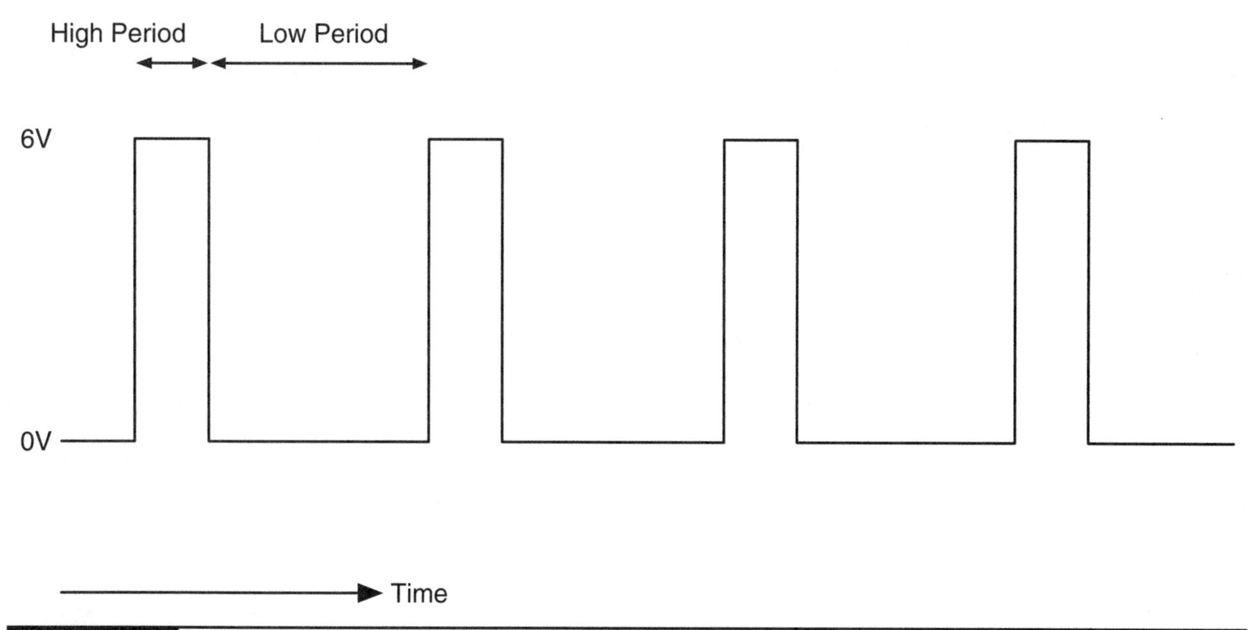

Figure 11-8 The servo motors

A pulse of 1.5 milliseconds sets the servo to its center position, a shorter pulse of 1.25 milliseconds to its leftmost position, and 1.75 to its rightmost position.

The servo motor expects there to be a pulse at least every 20 milliseconds for the servo to hold its position.

Fortunately for us, there is an Arduino library for driving servos that just allows us to use a command like this:

```
servo1.write(150);
```

This sets the angle of the servo between 0 and 180 degrees.

Summary

Content in his bath, the Evil Genius can now summon minions as required. In the next chapter, the Evil Genius will turn his attention to the taxing problem of automating his laundry.

CHAPTER 12

Delay Timer

WORLD DOMINATION IS a messy business and the battle to remove troublesome stains in the Evil Genius' laundry rages on, day after day. In fact, the Evil Genius even has to run his appliances at night, such is the demand for clean minion uniforms.

The Evil Genius therefore decided to create a device that would act as a delay timer, allowing him to run his clothes drier overnight (Figure 12-1).

This timer module can be used in a variety of ways, but the Evil Genius decided to modify his ancient clothes drier and incorporate the timer directly into it. You could, however, have it encased in a separate box if you desire.

After the appointed number of hours, the timer will close the contacts on its relay for 1/10th of a second. The relay contacts are wired to the contacts on the switch that the timer is simulating the pressing of.

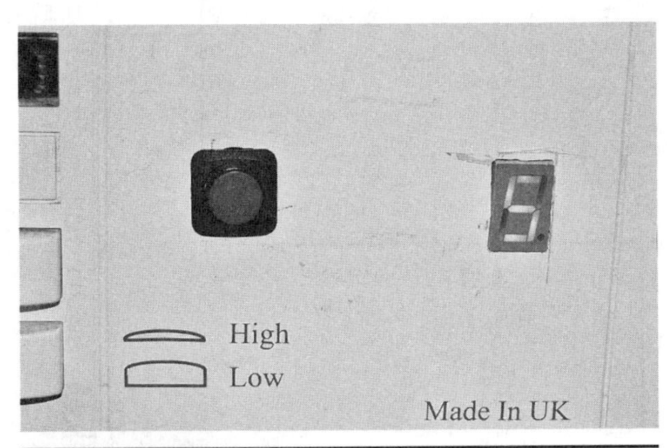

Figure 12-1 The delay timer

CAUTION If you intend this project to be attached to an AC-powered device, then be aware that you should only do this if you are absolutely sure you know what you are doing and are qualified to work with high-voltage equipment.

You should also be aware that modifying consumer equipment will void its warranty.

Construction

This is another "off-board" Arduino project, built onto some stripboard, and is a useful addition to the home automation projects we have been working on.

Figure 12-2 shows the schematic diagram for the project.

The design uses a common cathode seven-segment display to show the number of hours before the event is triggered—in this case, the clothes drier is started.

A single button cycles through the number of hours. If the button is not pressed for five seconds, the countdown begins.

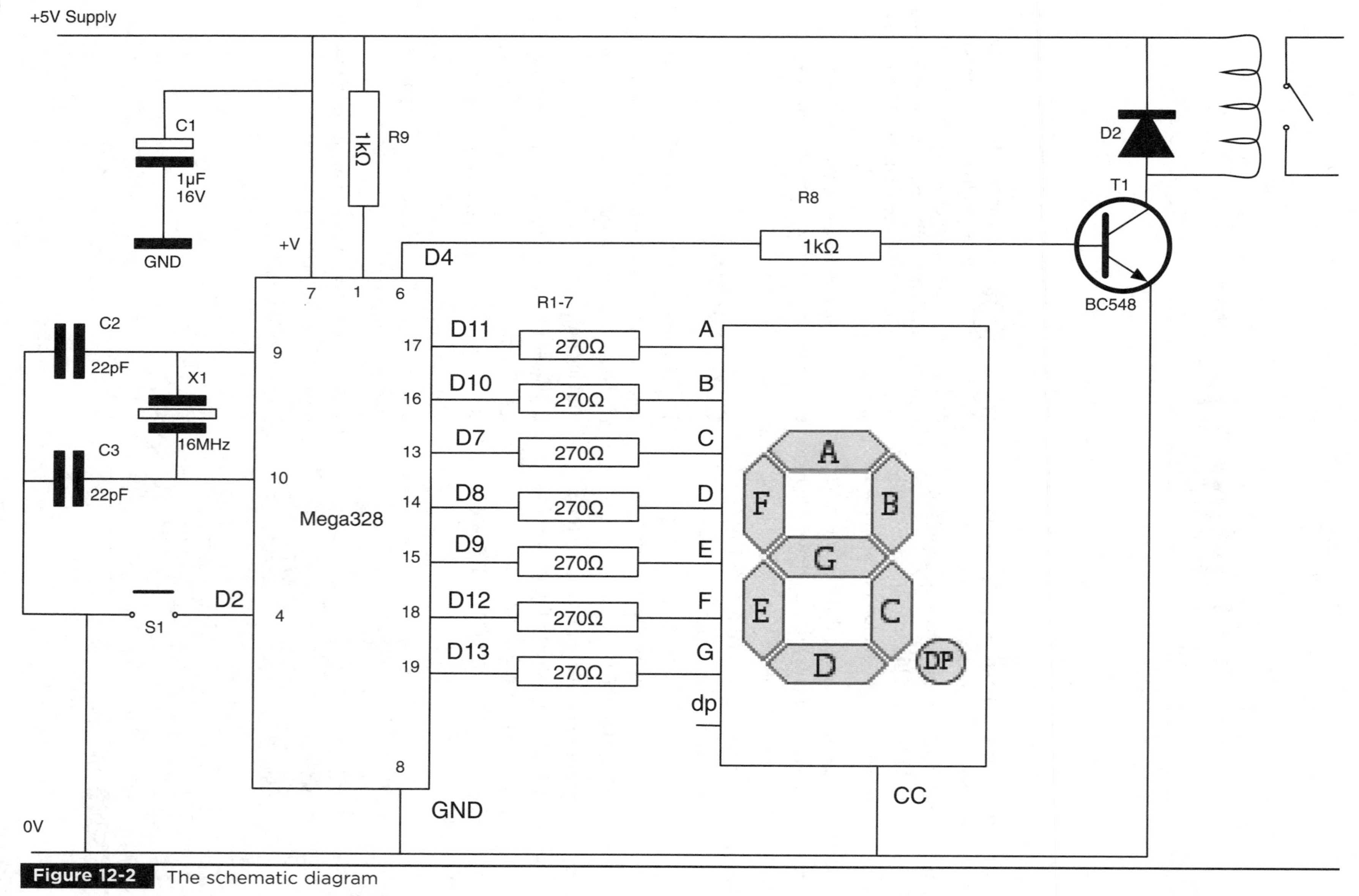

Figure 12-2 The schematic diagram

What You Will Need

You will need the following components listed in the Parts Bin to make the project.

This design uses the Arduino Uno to program the microcontroller. The official Arduino web site (www.arduino.cc) lists suppliers of the Uno. However, if you are on a budget, you can use a clone of the Arduino Uno.

The power supply is a recycled mobile phone charger. This needs to be a 5V type. Check this with a multimeter and also establish which wire in the lead is positive.

The relay used needs to be a 5V or 6V coil.

In addition to these components, you will also need the following tools listed in the Toolbox.

TOOLBOX

- An electric drill and assorted drill bits
- A hacksaw or Dremel rotary tool
- A multimeter
- A hot glue gun or epoxy glue
- A computer to program the Arduino
- A USB-type A-to-B lead

PARTS BIN

Part	Quantity	Description	Source
Microcontroller	1	ATMega328 with bootloader	SparkFun: DEV-10524
Relay	1	5 or 6V relay	Farnell: 1455502
T1	1	BC548	Farnell: 1467872
D1	1	Seven-segment common cathode LED—the 0.5-inch variety	Farnell: 1142439
D2	1	1N4001	Farnell: 1458986
C1	1	1μF electrolytic capacitor	Farnell: 1236655
C2, C3	2	22pF ceramic capacitor	Farnell: 1600966
R1-7	7	270Ω 0.5W metal film resistor	Farnell: 9340300
R8, R9	2	1kΩ 0.5W metal film resistor	Farnell: 9339779
X1	1	16-MHz crystal	Farnell: 1611761
Power supply	1	5V power supply	See description
Stripboard	1	28 strips of 18 holes	Farnell: 1201473
Header sockets	1	Two lengths of five-way socket cut from a socket header strip	Farnell: 1218869
IC socket	1	28-pin DIL IC socket	Farnell: 1824463
Switch	1	Push to make switch	Farnell: 1634627

Step 1. Prepare the Stripboard

Stripboard is a perforated board, with holes at 1/10-inch pitch. Behind the holes are strips of copper. Component leads are pushed through from the plain side and soldered to the copper strips. Figure 12-3 shows the stripboard for the sound link from the copper side.

Notice that some of the copper strips are cut where the IC is to be installed. These places are marked by an "X" on the stripboard layout in Figure 12-4. Such cuts in the strips are made by

Figure 12-3 The prepared stripboard

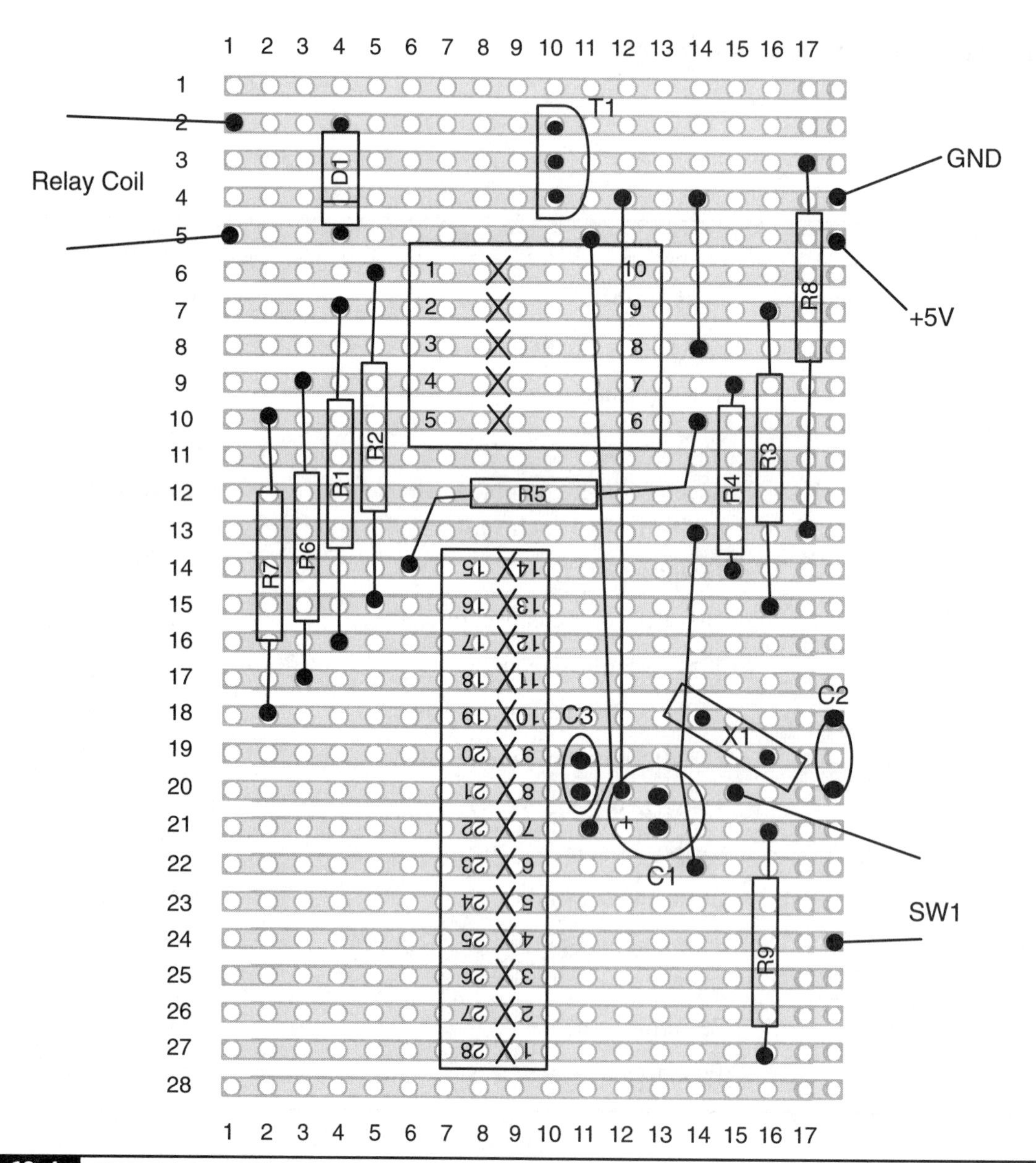

Figure 12-4 The stripboard layout

taking a drill bit, setting the tip into the hole where you want to make the cut, and twisting it between your fingers a few times to remove the copper track, without actually making much of a hole in the board itself.

So start by cutting a piece of stripboard that is comprised of 28 strips, each with 18 holes. You can do this with a strong pair of scissors, but it will usually result in some ragged edges. Greater neatness can be achieved by scoring the board with a craft knife and breaking it over the edge of a table. However, be careful when the stripboard breaks, it can leave sharp edges.

Using Figures 7-3 and 7-4 as a guide, make the 19 cuts in the track with a drill bit.

Step 2. Solder the Sockets

The next step is to solder the IC socket and the two strips of five-way socket header. The socket header can be cut from a long length to provide just five sockets using a craft knife (Figure 12-5).

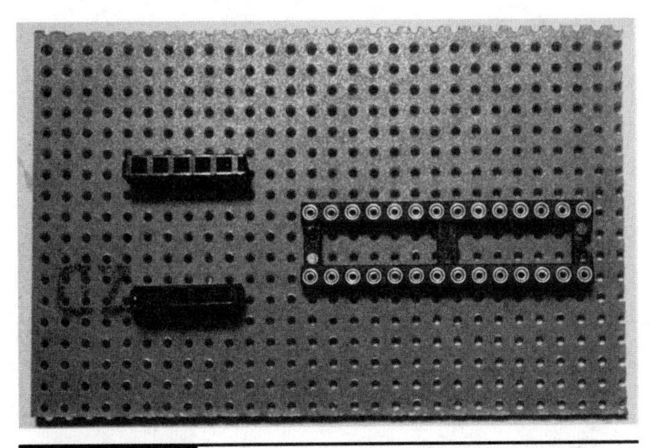

Figure 12-5 Soldering the sockets into place

The IC socket should have the end with a notch (indicating pin 1) at the outside (bottom) edge of the board.

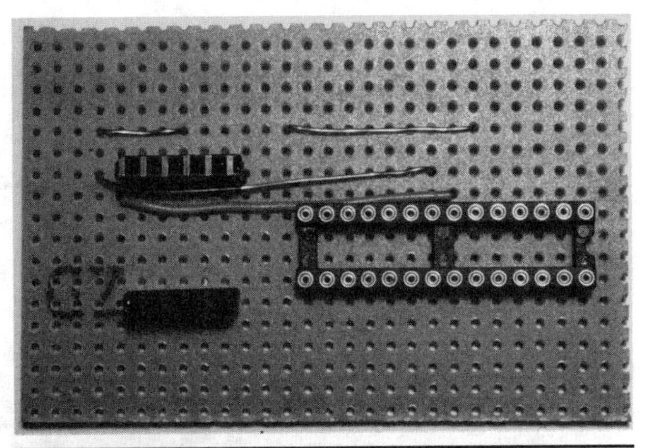

Figure 12-6 Soldering the links into place

Step 3. Solder the Links

There are just three links to be made (Figure 12-6). As you can see, the longer of these are left insulated to prevent accidental shorts.

Step 4. Solder the Remaining Components

We can now solder the rest of the components (Figure 12-7). Make sure that the electrolytic capacitor (C1) is the right way around (refer to Figure 12-4). The positive lead of the capacitor is normally longer and there should be a diamond symbol next to the negative lead.

Similarly make sure that the diode and transistor are both the correct way around.

Step 5. Install the Arduino Sketch

If you have built some of the earlier projects in this book, then chances are you will have already downloaded the zip file containing all the sketches from www.duinodroid.com. If you have not downloaded this, do so now. Unzip the file and move the whole Arduino Android folder to your sketches folder. In Windows, your sketches folder will be in My Documents/Arduino. On the Mac, you will find it in your home directory under Documents/Arduino/, and on Linux it will be in the sketchbook directory of your home directory.

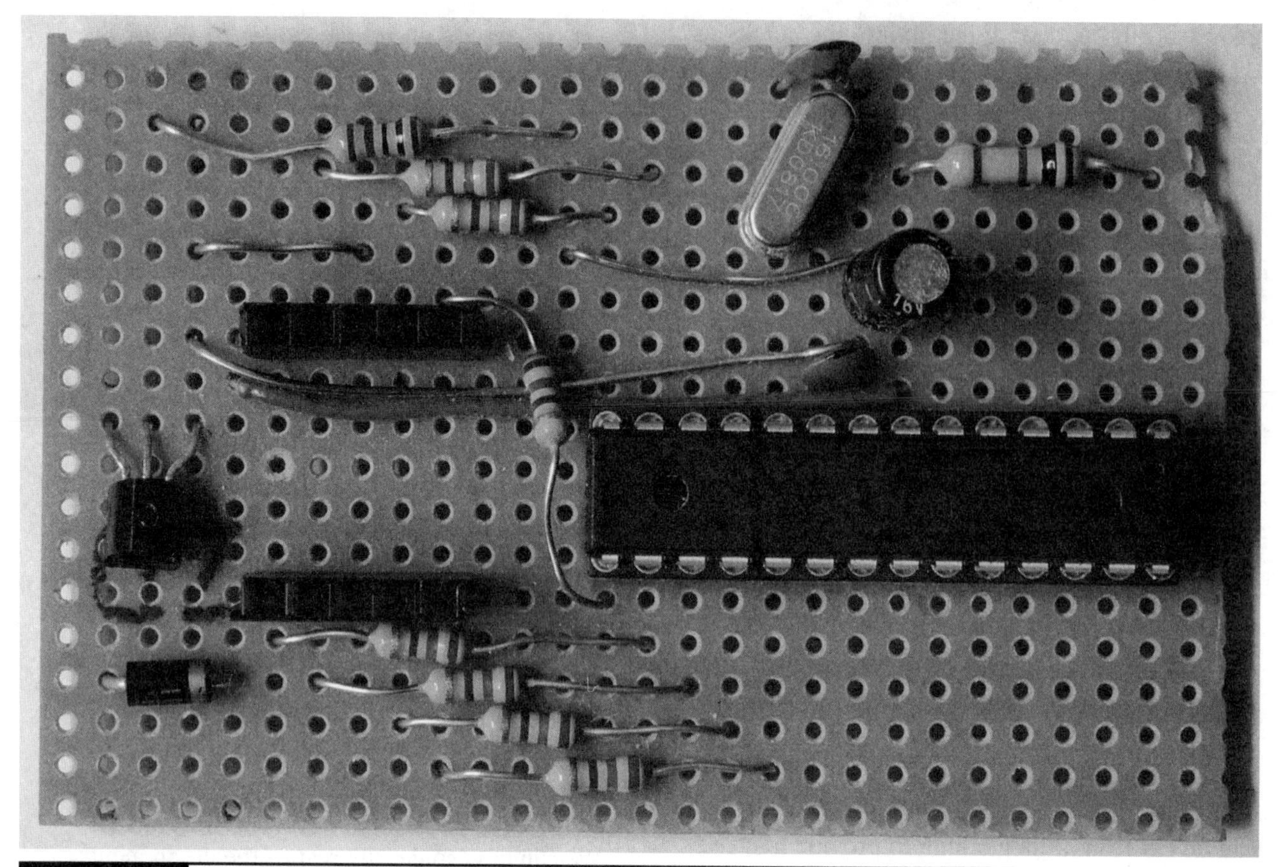

Figure 12-7 Soldering the remaining components

You will need to restart the Arduino software for it to pick up the new sketches.

From the File menu of the Arduino application, select Sketchbook, then Arduino Android, and then the sketch ch12_delay_timer.

Connect your Arduino board to your computer via USB.

We are now ready to upload the sketch to the board by clicking the "upload" icon (second to right on the toolbar). If you get an error message, you will need to check the type of board you are using and the connection.

Once the microcontroller is programmed, carefully remove the microcontroller chip and insert it into the IC socket on the stripboard. Make sure you get it the right way around. The notch on the socket should match the notch on the IC indicating the end of the chip with pin 1.

Step 6. Test

Now that we have completed the board, we can test it before we put it to use anywhere.

Solder leads between the board and the relay coil, and add short flying leads to the normally open connections of the relay. Also add short flying leads to connect to the switch and solder the leads from the power adaptor to the board (Figure 12-8).

Plug in the power adaptor. Initially, the LED will be blank. Momentarily touch the switch leads together and the LED should display the digit 3. This is the initial starting delay in hours. The Evil Genius usually likes to set the timer for at least three hours, and so he thought this was a more sensible default than one hour. Touching the leads together successively should increase the count displayed until it reaches 9, after which it should go to 1.

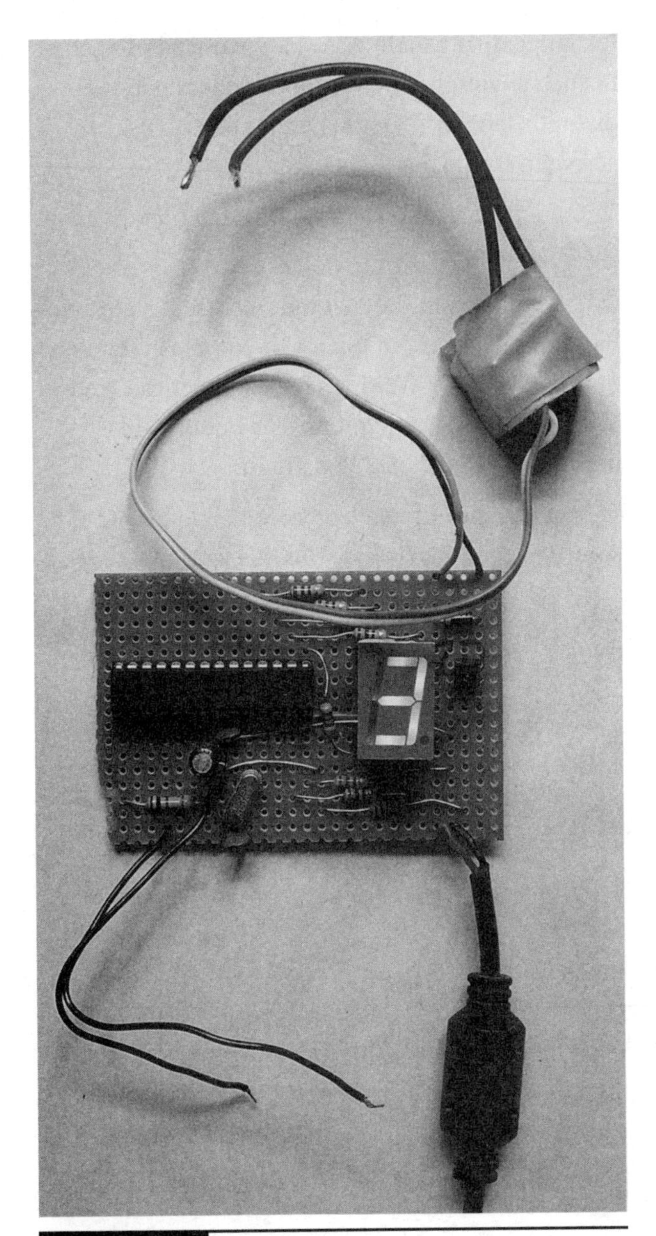

Figure 12-8 Testing the timer

A good way to test that the relay is functioning is to set the timer for one hour. It will immediately switch to a flashing "0" after the timer starts to indicate that the timer is at 0 hours and 59 minutes, before it will trigger.

Attach your multimeter with it set to "buzzer continuity mode," and then make a minion stand next to it for an hour to make sure that at the appointed time the relay clicks for a fraction of a second and the buzzer sounds.

Step 7. Installation

Please read the warning at the start of this chapter. AC electricity can kill you. So do not install this equipment unless you are absolutely sure you know what you are doing. Never work on an appliance while it is connected to the AC supply.

You can either attach the power adaptor to a separate mains socket, or add flying leads to the power adaptor as shown in Figure 12-9.

These are twisted tightly around the power pins, and then soldered. They should then be well covered with insulating tape. The flying leads can then be attached to a convenient AC supply within the appliance.

The Evil Genius was happy to drill a hole into the front of their appliance for the switch and cut out a rectangle for the display. The board was then fixed in place behind the front panel of the appliance (Figure 12-1).

Make sure that all the leads are secured using cable ties and are well away from anything that moves or gets hot.

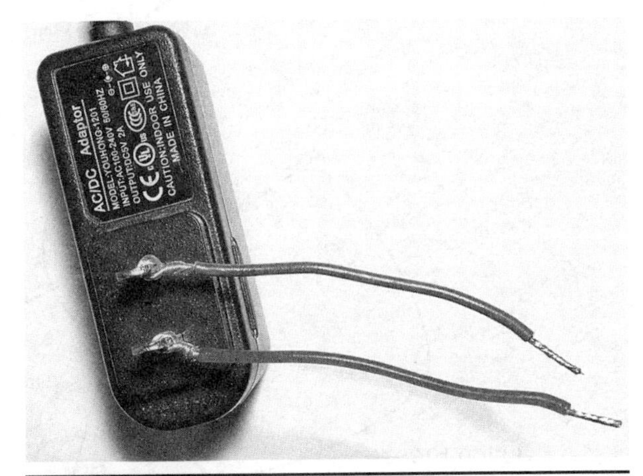

Figure 12-9 Attaching flying leads to the power adaptor

Theory

In this "Theory" section, we are going to look at software. We will look at how this system can be designed using a useful technique called state transition diagrams and then look at the sketch in some detail.

State Transition Diagrams

Although this timer is easy to operate, making a sketch that does things like start timing three seconds after the button was last pressed can be surprisingly difficult to get right. A useful design technique that usually saves a great deal of anguish in such situations is called the state transition diagram. The Evil Genius prefers not to use an acronym for this.

Figure 12-10 shows the state transition diagram for this timer.

The bubbles in the diagram represent states. So the timer is always in one of three states: Standby, Setting Delay, or Waiting (for the timer to expire). What happens when you press a button will depend on which state you are in.

You can move from one state to another when some "trigger" occurs. So, in this case, we start up

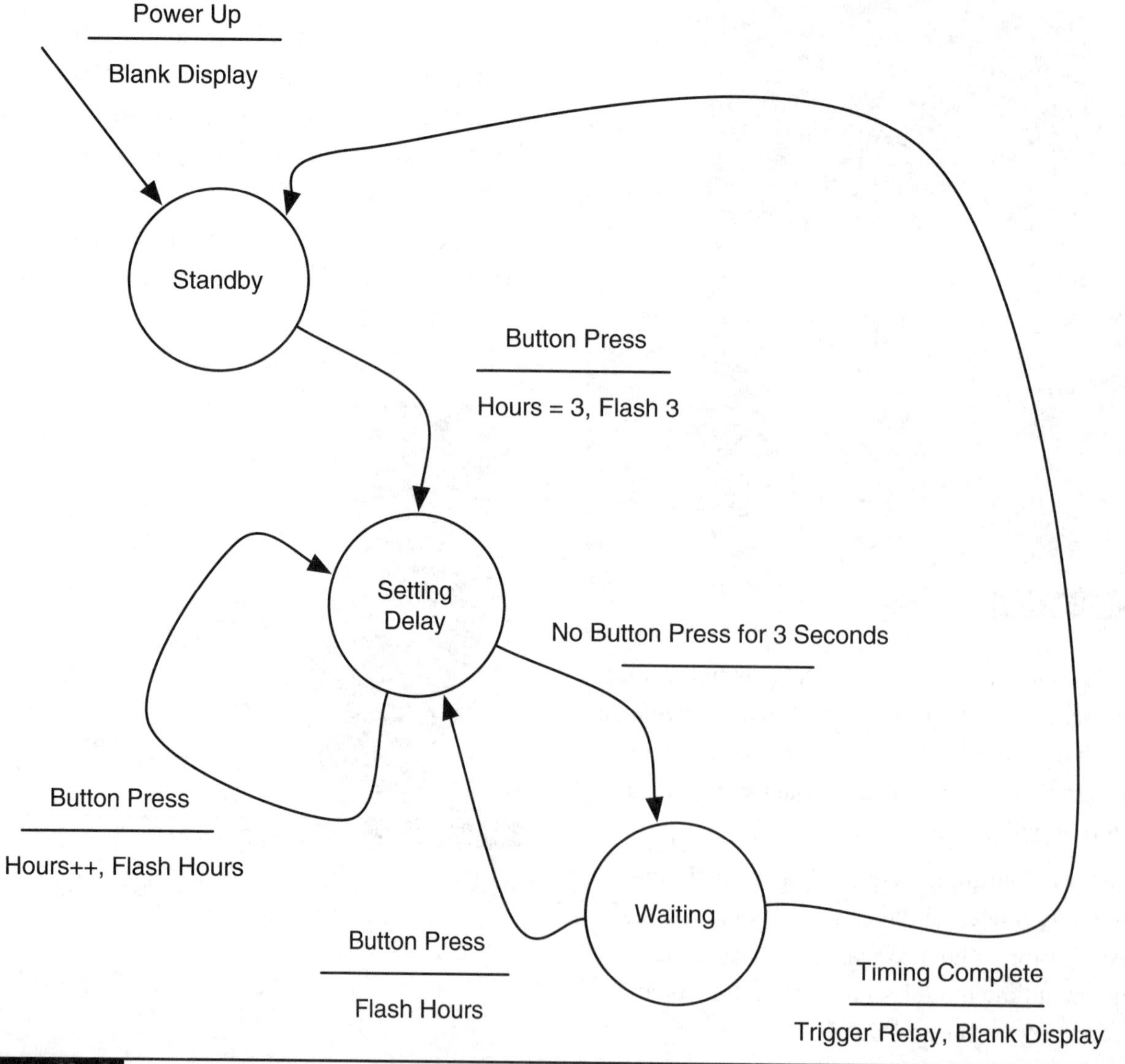

Figure 12-10 The state transition diagram

in the Standby state, but if the button is pressed (above the line in the label), then we perform the actions below the line, and then change to the state at the end of the line. In this case, the action is to set the "hours" variable to 3 and flash the digit. Pressing the button again when in the Setting Delay state will not result in a transition to another state, but will increment the "hours" variable.

We will only move to the Waiting state if there is no button press for three seconds. We will then stay in the Waiting state until either the time period has elapsed (in which case we return to Standby) or the button is pressed, in which case we return to the Setting Delay state.

The Arduino Sketch

As you read through this sketch, you will see how closely it mirrors the state transition diagram we just discussed.

```
int segmentPins[] = {11, 10, 7, 8, 9,
    12, 13};
int switchPin = 2;
int relayPin = 4;

byte digits[11][7] = {
//  a  b  c  d  e  f  g
  { 1, 1, 1, 1, 1, 1, 0},  // 0
  { 0, 1, 1, 0, 0, 0, 0},  // 1
  { 1, 1, 0, 1, 1, 0, 1},  // 2
  { 1, 1, 1, 1, 0, 0, 1},  // 3
  { 0, 1, 1, 0, 0, 1, 1},  // 4
  { 1, 0, 1, 1, 0, 1, 1},  // 5
  { 1, 0, 1, 1, 1, 1, 1},  // 6
  { 1, 1, 1, 0, 0, 0, 0},  // 7
  { 1, 1, 1, 1, 1, 1, 1},  // 8
  { 1, 1, 1, 1, 0, 1, 1},  // 9
  { 0, 0, 0, 0, 0, 0, 0}  // Blank
};

#define BLANK 10

#define STANDBY 0
#define SETTING_DELAY 1
#define WAITING 2
```

```
int state;
long hours;

void setup()
{
  for (int i = 0; i < 7; i++)
  {
    pinMode(segmentPins[i], OUTPUT);
  }
  pinMode(switchPin, INPUT);
  digitalWrite(switchPin, HIGH);
    // turn on pullup resistor
  pinMode(relayPin, OUTPUT);
  state = STANDBY;
  show(BLANK);
}

void loop()
{
  static long lastButtonPress = 0;
  static long startTime = 0;
  static long endTime = 0;
  boolean buttonPressed = !
    digitalRead(switchPin);
  if (state == STANDBY)
  {
    if (buttonPressed)
    {
      lastButtonPress = millis();
      state = SETTING_DELAY;
      hours = 3;
      flash(hours);
    }
  }
  else if (state == SETTING_DELAY)
  {
    if (buttonPressed)
    {
      lastButtonPress = millis();
      hours++;
      if (hours == 10)
      {
        hours = 1;
      }
      flash(hours);
    }
    else if ((millis() -
    lastButtonPress) > 3000)
    {
```

(continued)

```
        startTime = millis();
        endTime = startTime + (hours *
        60 * 60 * 1000);
        state = WAITING;
      }
    }
    else if (state == WAITING)
    {
      long timeNow = millis();
      hours = (endTime - timeNow) / 1000
      / 60 / 60;
      if (buttonPressed)
      {
        state = SETTING_DELAY;
        lastButtonPress = millis();
        flash(hours);
      }
      else if (hours < 0)
      {
        state = STANDBY;
        toggleRelay();
        show(BLANK);
      }
      else
      {
        flash(hours);
      }
    }
}

void show(int n)
{
  for (int i=0; i < 7; i++)
  {
    digitalWrite(segmentPins[i],
    digits[n][i]);
  }
}

void flash(int n)
{
  show(BLANK);
  show(n);
  delay(200);
  show(BLANK);
  delay(200);
  show(n);
}

void toggleRelay()
{
  digitalWrite(relayPin, HIGH);
  delay(100);
  digitalWrite(relayPin, LOW);
}
```

We start by defining an array of "ints" to contain the pins used by the segments of the display. We then define the pins to be used for the switch and the relay.

To display the segment patterns for the different digits, we use a two-dimensional array. We add an extra row to this array to represent all LEDs off. We will see later how this simplifies the code. In fact, it is the BLANK constant that refers to this row.

The next three constants (STANDBY, SETTING_DELAY, and WAITING) represent the three states we described in our state transition diagram. These could be any numbers, as long as they are different from each other.

We then declare two variables "hours" and "state". The variable "state" holds the current state that the timer is in, and "hours" is the number of hours set.

The "setup" function uses a loop to set all the segment pins as outputs and sets the modes of the other pins. Doing a "digitalWrite HIGH" to the switch pin enables the internal pull-up resistor for the pin.

We now come to the "loop" function, and it is here that we model the state transition diagram. We first define some static variables that will not be reset each time the function is run. These variables are used to record times at which events happen, such as the last time that a button was pressed.

We then have sections for each of the states. So, if we are in the STANDBY state, then we first record the time at which the button was pressed, then set the "state" variable to SETTING_DELAY, set the "hours" variable to 3, and then flash it.

You should be able to follow similar code for the other states and relate it to the state transition diagram.

Three other utility functions exist. The function "show" sets the appropriate segments for the digit supplied as its argument. The function "flash" does the same, but just flashes the digit for 200 milliseconds, and "toggleRelay" pulses the relay on for 100 milliseconds.

Summary

This is a useful little project that the Evil Genius can put to a number of good uses. It is also the final project-based chapter in this book.

I hope you have enjoyed building these projects, and will be inspired to combine your Android and Arduino devices in a suitably "Evil Genius" manner.

The following appendix provides you with a starting point for learning about Android Open Accessory development.

APPENDIX

Open Accessory Primer

Becoming adept at Android programming is a book in its own right. However, there is little more than the concise Google documentation available on the topic of the Open Accessory standard.

In this appendix, it is assumed you have a basic familiarity with Android and Arduino programming, but want to understand how the Open Accessory apps used in many of the projects work.

Learning Android Programming

There are many good books on the subject of learning Arduino. Buy one and work through at least the first few chapters to get an understanding of key concepts like *Activities* and *Intents*, as well as how to set up a development environment and use the Java programming language.

Many web resources and tutorials are available to get you started. However, the advantage of a well thought out and accurate book is that there will be a single consistent approach, whereas mixing and matching from web resources can be confusing to the beginner since each author will do things a different way.

Android development is a lot harder than Arduino programming—or rather there is a lot more to learn.

The author used the Eclipse plug-in for Android, which is a very popular approach and seems to work pretty well. Eclipse is an IDE like the Arduino IDE, but considerably more complex and powerful.

Arduino Programming

Learning Arduino is altogether a much easier task. Again, there are many good books out there and new ones appearing all the time. Look at the listings and reviews at your favorite book store. If you like the style of this book's author, you will find two books by him there: *Programming Arduino: Getting Started with Sketches* and *30 Arduino Projects for the Evil Genius*.

The Example

The Google documentation comes with a very comprehensive example for using the Open Accessory interface with an Arduino-based board called the ADK. This board was developed specifically to demonstrate the features of Open Accessory. It is based on an Arduino Mega with USB host facilities built in. These boards are not official Arduino boards and are more expensive and less common than the standard Arduino boards like the Uno.

The example that Google provides (using the ADK board) is rather too comprehensive. It is literally thousands of lines of code split across 18 classes. In this case, it's really not easy to see the forest for the trees.

The example we use in this Appendix pares this down to the bare essentials. This leaves you with a still-quite-substantial 300 lines of code in just two classes.

The purpose of the example (shown in Figure A-1) is to allow us to enter a number to be sent to the Arduino. The Arduino will then add one to the number and send it back. This is utterly pointless, but it does demonstrate communication in both directions.

The app also has the advantage that it will tell us everything that is going on in a "Log" area of the screen. This is useful because, unlike a regular app that can be connected to the Eclipse debugger so we can see what is going on, an Open Accessory app will be connected to an Arduino and not be available for USB debugging.

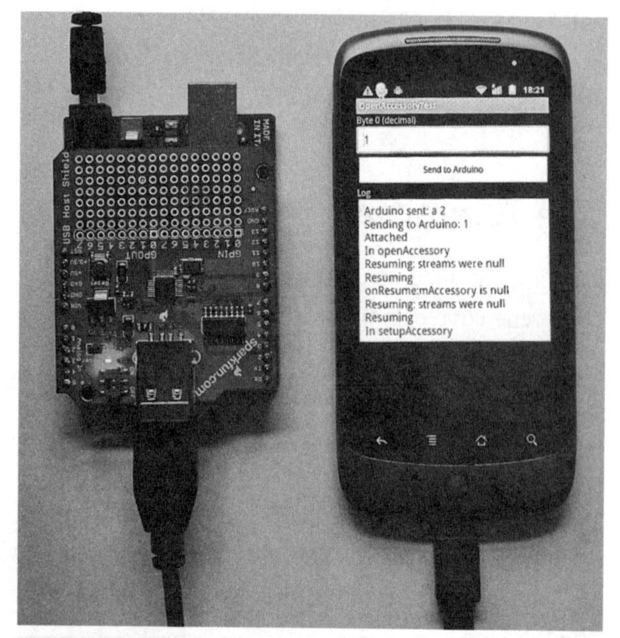

Figure A-1 The example app

On the Arduino

The Arduino part of this example is very straightforward. It is about one tenth of the size of the Android part.

Installing the Libraries

The Google Open Accessory requires two libraries to be installed into your Arduino environment. The first of these is a version of the USB host library, which is patched to work with standard Arduino hardware. This should be downloaded from microbridge.googlecode.com/files/usb_host_patched.zip.

If for any reason, you cannot find a download for any of the software used in these projects, please refer to this book's web site (www.duinodroid.com) where there will be instructions for obtaining the software elsewhere.

To install the library, download the zip file, unzip it, and move the unzipped folder to your Arduino libraries folder. In Windows, your libraries folder will be in My Documents/Arduino. On the Mac, you will find it in your home directory at Documents/Arduino/, and on Linux, it will be in the sketchbook directory of your home directory. If there is no "libraries" folder in your Arduino, you will have to create one. After installing the software, restart the Arduino software.

The second library, the AndroidAccessory library itself, is downloaded as part of the Adk package download, which can be found here: http://developer.android.com/guide/topics/usb/adk.html.

Click the link for "Adk package download." This will download a zip file. Unzip it and you will find that, inside a folder named ADK_release_0512, there are a couple of files and three folders. The only folder we are interested in is the one called firmware. This contains a folder named arduino_libs, and within that are two folders, each containing an Arduino library. One is the USB_Host_Shield, which we do not need to install because we have already just installed a version of it. However, we do need to install the AndroidAccessory library.

To do this, just move the whole AndroidAccessory folder to your "libraries" folder the same way you did the USB host library.

You will need to restart the Arduino software for it to pick up the new libraries.

The Sketch

The sketch is listed here in full. It is also included with the zip file of project sketches that can be downloaded from the book's web site (www.duinodroid.com). The sketch is called apA_open_accessory_test.

```
#include <Max3421e.h>
#include <Usb.h>
#include <AndroidAccessory.h>

AndroidAccessory acc("Simon Monk",
      "OpenAccessoryTest",
      "DemoKit Arduino Board",
      "1.0",
      "http://www.duinodroid.com",
      "0000000012345678");

void setup()
{
  acc.powerOn();
}

void loop()
{
  byte msg[1];
  if (acc.isConnected())
  {
    int len = acc.read(msg,
    sizeof(msg), 1);
    if (len >= 1)
    {
      byte value = msg[0];
      sendMessage(value + 1);
    }
  }
}

void sendMessage(int value)
{
  if (acc.isConnected())
  {
    byte msg[2];
    msg[0] = value >> 8;
    msg[1] = value & 0xff;
    acc.write(msg, 2);
  }
}
```

After the first #include statements, the first real code creates a new instance of the class AndroidAccessory. The arguments to the constructor tell your Android device all about the accessory that has just been connected to it.

The first three parameters are the "vendor", the "application name", and the version—in this case, "Simon Monk", "OpenAccessoryTest", and "1.0". These will be used by the Android side of the interface to automatically start the appropriate app on the Android phone when the accessory is connected.

The next argument is a URL, and if the appropriate app is not installed, the Android phone will present this and allow direct navigation to that URL, opening it in a browser. The app can then be downloaded, installed, and run.

The final argument is a serial number for the app.

The setup method only needs to invoke the powerOn method, which starts the USB communication.

In the loop function, we use isConnected to determine if there is a connection to the Android device. Note that if it is not connected, it will attempt to make a connection. So if the devices become disconnected, there is no need to run powerOn again or reset the Arduino.

The read method determines if there is a message waiting to be read. If there is, it returns the size of the message in bytes. If there is a message, then it is assumed to be a single byte. Once read, the byte is incremented and then sent to the function "sendMessage" to return the new value to the Android device.

The sendMessage function first checks to see if there is a connection. If there is or one can be opened when isConnected is called, then a message is constructed into a byte array of two bytes. This is sending an "int" back to the Android device. If you want to send other data back, they must be packed into a byte array in this manner.

Finally, the write method is supplied with the byte array and message length that will be sent back to the listening Android device.

Android

The code for this side of the link is too copious to list in full here. We will list it all in parts, but you will find this explanation easier if you fetch the source code for it from www.duinodroid.com and have it in a text editor or IDE while it is discussed.

Auto-start and Download

One of the best features of the Open Accessory standard is that your Arduino can effectively tell your Android phone what app it needs and cause the app to automatically start when the device is connected. It can even direct the phone to a URL, from which the app can be downloaded.

The Arduino side of this is defined in the constructor of the AndroidAccessory.

```
AndroidAccessory acc("Simon Monk",
      "OpenAccessoryTest",
      "DemoKit Arduino Board",
      "1.0",
      "http://www.duinodroid.com",
      "0000000012345678");
```

The corresponding parts of the Android solution are spread across two files.

First, you must add the following lines to the Android manifest for the project. Inside the "activity" tag, add the tag:

```
<intent-filter>
  <action android:name="android
    .hardware.usb.action.USB_ACCESSORY
    _ATTACHED"/>
</intent-filter>
```

Inside the "application" tag, add the following tags:

```
<uses-library android:name="com.android
    .future.usb.accessory"/>
<meta-data android:name="android.hardware
    .usb.action.USB_ACCESSORY_ATTACHED"
  android:resource="@xml/accessory
    _filter"/>
```

The second tag names the file xml/accessory _filter, and it is this file that is used to match against when deciding if this app is being asked to start.

```
<?xml version="1.0" encoding="utf-8"?>

<resources>
    <usb-accessory manufacturer=
    "Simon Monk"
    model="OpenAccessoryTest"
    version="1.0" />
</resources>
```

Lifecycle

The Android app has to cope with a number of different situations and events relating to both the accessory and the general lifecycle of the main Activity for the app.

Figure A-2 shows the lifecycle of an Activity in Android, along with the methods that get called each time the Activity changes from one state to another.

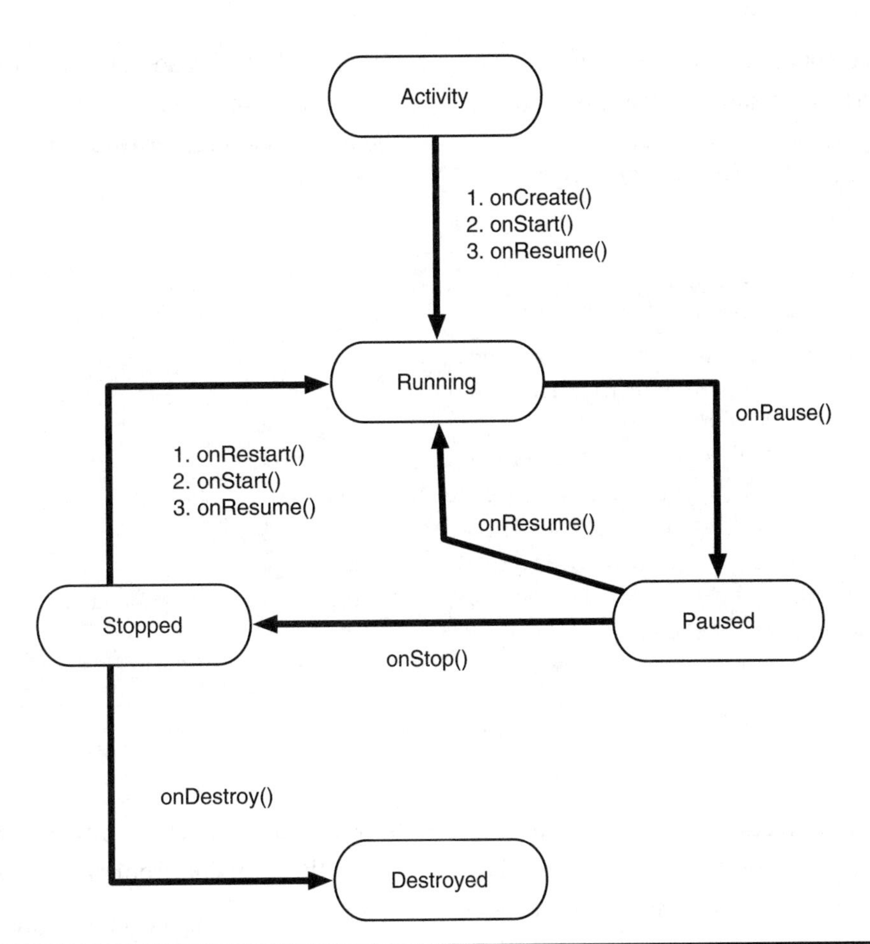

Figure A-2 An Android Activity lifecycle

Opening the Accessory

When an Activity is first created, three functions will be called: onCreate, onStart, and onResume. In our example Activity, we implement onCreate and onResume, as shown below.

The onCreate method is very like the onCreate of any Activity. It sets member variables for the various controls on the Activity and creates a listener to be fired when the Send button is pressed. It also calls setupAccessory.

```
public void onCreate(Bundle savedInstanceState) {
    super.onCreate(savedInstanceState);
    setContentView(R.layout.main);
    mByteField = (EditText) findViewById(R.id.messagebyte);
    mResponseField = (EditText) findViewById(R.id.arduinoresponse);
    mSendButton = (Button) findViewById(R.id.sendButton);
    mSendButton.setOnClickListener(new OnClickListener() {
      public void onClick(View v) {
        sendMessageToArduino();
      }
    });
    setupAccessory();
  }
```

The setupAccessory method creates an instance of UsbManager and associates it with the Activity. It then registers the ACTION_USB_ACCESSORY_DETACHED action with the broadcast receiver mUsbReceiver so it can invoke the closeAccessory method from the broadcast receiver when the accessory is detached.

```
private void setupAccessory() {
      log("In setupAccessory");
      mUsbManager = UsbManager.getInstance(this);
      mPermissionIntent = PendingIntent.getBroadcast(this, 0, new Intent(
          ACTION_USB), 0);
      IntentFilter filter = new IntentFilter(ACTION_USB);
      filter.addAction(UsbManager.ACTION_USB_ACCESSORY_DETACHED);
      registerReceiver(mUsbReceiver, filter);
      if (getLastNonConfigurationInstance() != null) {
        mAccessory = (UsbAccessory) getLastNonConfigurationInstance();
        openAccessory(mAccessory);
      }
  }
```

The function setupAccessory will always be called following a call to onResume. So, it will be possible to retrieve a handle to mAccessory using the getLastNonConfigurationInstance method. It then opens the accessory. If mAccessory has not been instantiated yet, it will be created in a later call to this function from onResume.

The openAccessory method creates input and output streams to the accessory, as well as starts a separate thread that will be used to listen for incoming messages.

```
private void openAccessory(UsbAccessory accessory) {
      log("In openAccessory");
      mFileDescriptor = mUsbManager.openAccessory(accessory);
      if (mFileDescriptor != null) {
        mAccessory = accessory;
        FileDescriptor fd = mFileDescriptor.getFileDescriptor();
        mInputStream = new FileInputStream(fd);
        mOutputStream = new FileOutputStream(fd);
        Thread thread = new Thread(null, this, "OpenAccessoryTest");
        thread.start();
        alert("openAccessory: Accessory opened");
        log("Attached");
      } else {
        log("openAccessory: accessory open failed");
      }
  }
```

As we mentioned earlier, when an Activity starts, it calls onCreate and onResume.

```
public void onResume() {
    log("Resuming");
    super.onResume();

    if (mInputStream != null &&
    mOutputStream != null) {
      log("Resuming: streams were not null");
    } else {
      log("Resuming: streams were null");
      establishPermissionsAndOpen
      Accessory();
    }
  }
```

The onResume method tests to see if the streams are null. If they are, then it calls the method establishPermissionsAndOpenAccessory. As the name suggests, this function ensures that the Activity has been granted the necessary permissions and then opens the streams to the accessory by calling the openAccessory method again, shown below.

```
private void establishPermissionsAndOpenAccessory() {
    UsbAccessory[] accessories = mUsbManager.getAccessoryList();
    UsbAccessory accessory = (accessories == null ? null : accessories[0]);
    if (accessory != null) {
      if (mUsbManager.hasPermission(accessory)) {
        openAccessory(accessory);
      } else {
       synchronized (mUsbReceiver) {
          if (!mPermissionRequestPending) {
            mUsbManager.requestPermission(accessory, mPermissionIntent);
            mPermissionRequestPending = true;
          }
        }
      }
    } else {
      log("establishPermissionsAndOpenAccessory:mAccessory is null");
    }
  }
```

Broadcast Receiver

A broadcast receiver allows an app to intercept system announcements. In this case, we want to know when someone has unplugged the accessory. An instance of Broadcast Receiver is created and assigned to the member variable mUsbReceiver. See the code listing on the next page.

```
private final BroadcastReceiver mUsbReceiver = new BroadcastReceiver() {
    @Override
    public void onReceive(Context context, Intent intent) {
      String action = intent.getAction();
      if (UsbManager.ACTION_USB_ACCESSORY_DETACHED.equals(action)) {
        UsbAccessory accessory = UsbManager.getAccessory(intent);
        if (accessory != null && accessory.equals(mAccessory)) {
          log("Detached");
          closeAccessory();
        }
      }
    }
  };
```

This receiver is registered as a receiver in the setupAccessory method that is called from onCreate.

In the event that the accessory is disconnected from the phone, the closeAccessory method will be called.

```
private void closeAccessory() {
    log("In closeAccessory");
    try {
      if (mFileDescriptor != null) {
        mFileDescriptor.close();
      }
    } catch (IOException e) {
    } finally {
      mFileDescriptor = null;
      mAccessory = null;
      mInputStream = null;
      mOutputStream = null;
    }
  }
```

Sending Data

When it comes to sending data from Android to the Arduino, we are back in the familiar territory of writing to a stream.

In this example, we are just sending a single byte to the Arduino. The method sendMessageToArduino is invoked when we press the "Send to Arduino" button. It retrieves the text from the input field and validates it as being a number and then passes the byte value to sendCommand.

```
public void sendMessageToArduino() {
    String valueStr =
    mByteField.getText() .toString();
    byte val;
    try {
      val = Byte.parseByte(valueStr);
      log("Sending to Arduino: " + val);
      sendCommand(val);
    } catch (NumberFormatException e) {
      // TODO Auto-generated catch block
      e.printStackTrace();
      alert("The Byte should be a
      number between 0 and 255");
    }

  }
```

Send command constructs a byte array (of just a single byte) and writes it to the output stream. The byte array is not necessary in this case, but if you were to pass more data, you would just construct a byte array of the size you needed, pack it with the data and then write it to the stream.

```
public void sendCommand(byte value) {
  byte[] buffer = new byte[1];
  buffer[0] = (byte) value;
  if (mOutputStream != null) {
    try {
      mOutputStream.write(buffer);
    } catch (IOException e) {
      log("Send failed: " +
      e.getMessage());
    }
  } else {
    log("Send failed: mOutStream was
    null");
  }
}
```

Receiving Data

We cannot allow the main user interface thread of the Activity to block and wait for a response from the Arduino. Indeed, it may be the Arduino that will be initiating the communication. For this reason, the openAccessory method starts a separate thread that listens for incoming messages from the Arduino.

```
public void run() {
    int ret = 0;
    byte[] buffer = new byte[16384];
    int i;

    while (true) {
      try {
        ret = mInputStream.read(buffer);
      } catch (IOException e) {
        break;
      }

      i = 0;
      while (i < ret) {
        int len = ret - i;
        if (len >= 2) {
          Message m = Message.obtain
          (mHandler);
          int value = composeInt
          (buffer[i], buffer[i + 1]);
          m.obj = new ValueMsg('a',
          value);
          mHandler.sendMessage(m);
        }
        i += 2;
      }
    }
}
```

The thread loops forever, waiting to read something on the input stream. When it finds a message (when read returns a value greater than zero), it constructs an int from the next two bytes in the message using the composeInt utility method.

```
private int composeInt(byte hi, byte lo) {
    int val = (int) hi & 0xff;
    val *= 256;
    val += (int) lo & 0xff;
    return val;
}
```

Having received a message from the Arduino, we now need to display it. But to be able to do this, we need to interact with the Activity, which requires the use of a Handler.

```
Handler mHandler = new Handler() {
    @Override
    public void handleMessage(Message
    msg) {
      ValueMsg t = (ValueMsg) msg.obj;
      log("Arduino sent: " +
      t.getFlag()
      + " " + t.getReading());
    }
};
```

The handler is then sent a message that it can add to the log field. The message is encapsulated in a class called ValueMsg. This is rather over-engineered for this application, but it is a good mechanism when receiving more structured information from the Arduino such as sensor readings. The ValueMsg class has members for an int reading and a flag that could be used to indicate the type of reading.

In a more practical example, this class would be extended to include any additional data in the Arduino message.

```
public class ValueMsg {
  private char flag;
  private int reading;

  public ValueMsg(char flag, int
   reading) {
    this.flag = flag;
    this.reading = reading;
  }

  public int getReading() {
    return reading;
  }

  public char getFlag() {
    return flag;
  }
}
```

Finally, the method "log" adds a line to the top of the log field in the app, and "alert" displays a message in a dialog.

Conclusion

The Arduino side of Open Accessory is pretty easy to use; however, the Android side is considerably more complex.

When it comes time to build your own Open Accessory app, a good starting point is to take the example project here, which can be downloaded from www.duinodroid.com, and modify it to suit your requirements.

You may also find a project in this book that is close to what you are trying to do and modify the app to suit your needs. These projects are all based on the model used in the ADK example app provided by Google.

Index

References to figures are in italics.

A

ADK, 17, 18, 183
  See also Open Accessory
Adk package download, 184
Amarino, 12, 13
Android
  Activity lifecycle, 186–187
  broadcast receivers, 189–190
  Open Accessory on, 186–192
  programming, 183
  receiving data, 191–192
  sending data to the Arduino, 190–191
  software for home automation controller, 99–101
Android apps, see *individual projects*
Android Geiger counter. *See* Geiger counter
Android light show. *See* light show
Arduino
  Duemilanove, 89
  Open Accessory on, 184–186
  programming, 183
  Uno, 4, 89
  See also sketches

B

Beeper, 104–106
Bluetooth modules, 4–5, 6–7
Bluetooth robot
  Android app, 15
  Arduino sketch, 14–15
  attaching pin headers to shield, 5–6
  attaching screw terminals to shield, 6
  construction, 3–14
  cutting the case bottom and fixing the castor, 8
  final wiring, 8–9
  fixing motors and battery box to case, 7–8
  installing Bluetooth module, 6–7
  installing the Android app, 12–13
  installing the real Arduino sketch, 12
  overview, 3
  Parts Bin, 5
  schematic diagram, *4*
  testing the motors, 10–12
  theory, 14–15
  Toolbox, 5
  trying out the robot, 13–14
broadcast receivers, 189–190

C

controller. *See* home automation controller

D

delay timer
  construction, 171–177
  installation, 177
  installing the Arduino sketch, 175–176

delay timer *(continued)*
- overview, 171
- Parts Bin, 173
- preparing the stripboard, 174–175
- schematic diagram, *172*
- sketch, 179–181
- soldering remaining components, 175
- soldering the links, 175
- soldering the sockets, 175
- state transition diagrams, 178–179
- stripboard layout, *174*
- testing, 176–177
- theory, 178–181
- Toolbox, 173

door lock. *See* RFID door lock

E

Eclipse plug-in, 183
encoding data as sound, 103–106

G

gear motors, 5
Geiger counter
- Android app, 34–35
- attaching low-lying components, 22
- attaching pin headers to shield, 21–22
- boxing the project, 27–29
- construction, 18–29
- final wiring, 24
- installing the Android app, 26
- installing the Arduino sketch, 25
- installing the GM tube, 26
- installing the Open Accessory libraries, 24–25
- overview, 17
- Parts Bin, 20
- schematic diagram, *19*
- soldering remaining components, 22–23
- soldering the leads to the Arduino pins, 23–24
- testing, 27
- testing the high-voltage supply, 25–26
- theory, 30–35
- Toolbox, 21

Geiger-Müller (GM) tube, 17, 26, 30
GM tube. *See* Geiger-Müller (GM) tube

H

home automation controller
- adding the power control to, 120–123
- Android software, 99–101
- attaching trailing leads, 93–94
- boxing the controller, 98–99
- decoding the sounds on the Arduino, 108–109
- encoding data as sound, 103–106
- Internet access, 102–103
- modifying, 136–138
- overview, 85–87
- Parts Bin, 87–89
- preparing the stripboard, 90–91
- schematic diagram, *88*, *106*
- soldering remaining components, 92–93
- soldering the IC, 92
- soldering the links, 91
- soldering the resistors and diode, 91–92
- sound interface electronics, 106–108
- sound link module, 87–99
- testing, 94–98
- theory, 103–109
- Toolbox, 90
- *See also* delay timer; power control; RFID door lock; signaling flags; smart thermostat

I

infrared remote controls, 60–61
- *See also* TV remote

L

libraries, installing, 184–185
light show
- Android app, 53
- Arduino sketch, 51–53
- attaching link wires for the USB shield, 40
- attaching socket, screw terminal and transistor, 45
- boxing the project, 48–49
- connecting everything together, 47
- connecting the 5V supply, 41–42
- connecting the switch and power leads, 42
- constructing the Droid Accessory Base, 38–43
- constructing the light show project, 43–49
- cutting perfboard to size, 44–45

fitting the crystal and other components, 41
fitting the remaining links, 41
installing the Android app, 48
installing the Arduino sketch, 48
MOSFETs, 50
overview, 37–38
Parts Bin, 39, 44
Pulse Width Modulation, 50–51
schematic diagram, *39*, *43*
soldering remaining components, 46–47
soldering the IC socket in place, 40
testing, 42–43, 48
theory, 50–53
Toolbox, 38, 44
using the project, 50
lock. *See RFID door lock*

M

MOSFETs, 50
motor shield, 5

O

Open Accessory, 18
on the Android, 186–192
Android and Arduino programming, 183
on the Arduino, 184–186
example, 183–184
installing the libraries, 184–185
sketch, 185–186
Open Accessory development kit. *See* ADK

P

Pachube
accounts, 68–69
web site, 63–64
perfboard, 112
cutting to size, 44–45
Polulu, 5
power control
adding it to the home automation controller, 120–123
attaching leads to the remote control PCB, 115–117
constructing the power control module, 112–120
disassembling the remote control, 114–115
electronics, 111–112
overview, 111
Parts Bin, 114
perfboard layout, *118*
placing components on the perfboard, 117–119
relays, 124–125
schematic diagram, 113
setting up your home, 124
sketches, 125–128
soldering the connections, 119
testing, 119–120
theory, 124–128
Toolbox, 114
wiring diagram with sound interface, *121*
programming
Android, 183
Arduino, 183
Pulse Width Modulation, 50–51
PWM. *See* Pulse Width Modulation

R

relays, 124–125
RFID door lock
boxing the project, 154–155
connecting everything together, 152, *153*
construction, 146–156
installation, 156
multicolor flashes, 156–157
overview, 145–146
Parts Bin, 146–148
preparing the stripboard, 149–150
programming and installing the microcontroller, 152
schematic diagram, *147*
soldering remaining components, 151–152
soldering the IC socket and switches, 150–151
soldering the links, 150
soldering the resistors and diode, 150
stripboard layout, *149*
testing, 152–154
theory, 157–161

RFID door lock *(continued)*
Toolbox, 149
using the system, 156–157
robot. *See* Bluetooth robot

S

servos, 168–169
signaling flags
constructing the wooden platform, 165–166
construction, 164–168
fixing the Arduino to the base, 167
making and attaching the flags, 168
overview, 163–164
Parts Bin, 164
programming the Arduino, 167, *168*
schematic diagram, *166*
servos, 168–169
testing, 167
theory, 168–169
Toolbox, 165
wiring diagram, *166*
wiring the servos to the terminal blocks, 166–167
sketches
for the Bluetooth robot, 14–15
for the delay timer, 179–181
for the Geiger counter, 30–34
for the home automation controller, 108–109
for the light show, 51–53
Open Accessory, 185–186
for the power control, 125–128
for the RFID door lock, 157–161
for the smart thermostat, 141–144
smart thermostat
boxing the project, 139–140
connecting everything together, 135–136
construction, 130–140
installation, 140
modified perfboard layout, *137*
modifying the home automation controller, 136–138
one-wire sensors, 141
overview, 129–130
Parts Bin, 132
preparing the stripboard, 133–134
programming and installing the microcontroller, 135
schematic diagram, *131*
sketch, 141–144
soldering remaining components, 135
soldering the links and IC socket, 134
soldering the resistors and diode, 134
stripboard layout, *133*
testing, 138
theory, 141–144
Toolbox, 133
using the system, 140–141
sound, encoding data as, 103–106
sound interface electronics, 106–108
Sparkfun, 5
state transition diagrams, 178–179
stripboard, 90–91

T

temperature logger
attaching components to the screw terminal, 65–66
boxing the project, 67–68
construction, 64–68
installing the Android app, 66
installing the Arduino sketch, 66
making a Droid Accessory Base, 65
overview, 63
Parts Bin, 65
schematic diagram, *64*
testing, 67
theory, 70–71
Toolbox, 65
using the project, 68–70
thermostat. *See* smart thermostat
timer. *See* delay timer
TV remote
boxing the project, 59–60
construction, 56–60
cutting stripboard to size, 57
installing the Android app, 59
installing the Arduino sketch, 58–59
IR remote controls, 60–61
making a Droid Accessory Base, 57
overview, 55

Parts Bin, 56
schematic diagram, *56*
soldering components, 57–58
testing, 59
theory, 60–61
Toolbox, 57
using the project, 60
wiring, 58

U

ultrasonic range finder
boxing the project, 77–78
construction, 74–78
installing the Android app, 76
installing the Arduino sketch, 76
making a Droid Accessory Base, 75
overview, 73–74
Parts Bin, 75
schematic diagram, *74*
soldering the laser and resistor, 76
testing, 77
theory, 79–81
Toolbox, 75
using the project, 79
ultrasonic range finding, 79–81

W

wire, 89